THE GRAND PRIX YEAR

PHILLIP HORTON

THE GRAND PRIX YEAR

AN INSIDER'S GUIDE TO FORMULA 1 RACING

BLOOMSBURY SPORT
LONDON · OXFORD · NEW YORK · NEW DELHI · SYDNEY

1	Australia	**13**	Belgium
2	China	**14**	Hungary
3	Japan	**15**	The Netherlands
4	Bahrain	**16**	Italy
5	Saudi Arabia	**17**	Azerbaijan
6	Miami	**18**	Singapore
7	Emilia Romagna	**19**	United States
8	Monaco	**20**	Mexico City
9	Spain	**21**	São Paulo
10	Canada	**22**	Las Vegas
11	Austria	**23**	Qatar
12	Great Britain	**24**	Abu Dhabi

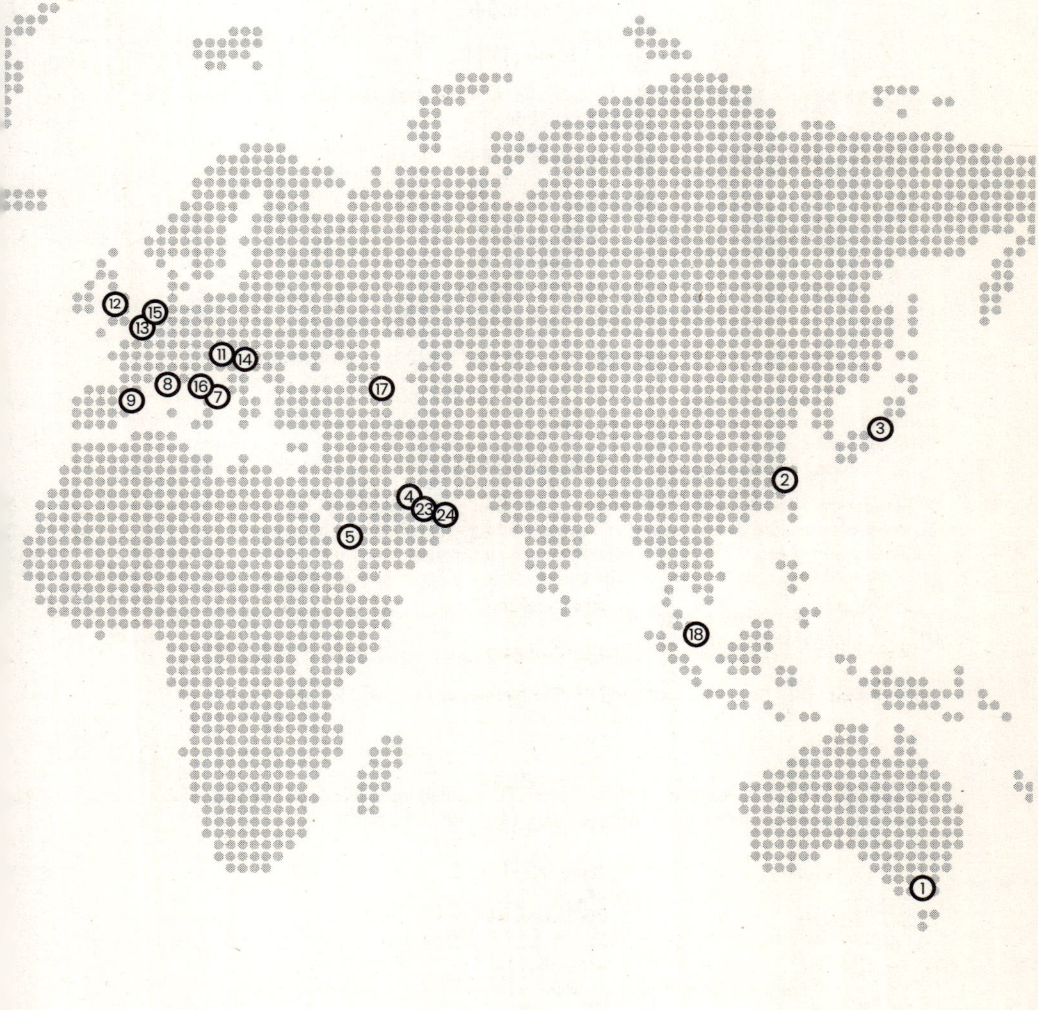

12
15
13
11
14
8
16
7
9
17
3
2
4
23
24
5
18
1

BLOOMSBURY SPORT
Bloomsbury Publishing Plc
50 Bedford Square, London, WC1B 3DP, UK
29 Earlsfort Terrace, Dublin 2, Ireland

BLOOMSBURY, BLOOMSBURY SPORT and the Diana logo are trademarks of Bloomsbury Publishing Plc

First published in Great Britain 2025

Copyright © Phillip Horton, 2025

Inside photography © Phillip Horton, with the exception of p.43
© Clive Mason - Formula 1 / Contributor, Getty Images

Phillip Horton has asserted his right under the Copyright, Designs and Patents Act, 1988, to be identified as Author of this work

For legal purposes the Acknowledgements on p. 254 constitute an extension of this copyright page

All rights reserved. No part of this publication may be reproduced or transmitted in any form or by any means, electronic or mechanical, including photocopying, recording, or any information storage or retrieval system, without prior permission in writing from the publishers

Bloomsbury Publishing Plc does not have any control over, or responsibility for, any third-party websites referred to or in this book. All internet addresses given in this book were correct at the time of going to press. The author and publisher regret any inconvenience caused if addresses have changed or sites have ceased to exist, but can accept no responsibility for any such changes

A catalogue record for this book is available from the British Library

Library of Congress Cataloguing-in-Publication data has been applied for

ISBN: HB: 978-1-3994-1489-0; TPB: 978-1-3994-2043-3; ePUB: 978-1-3994-1490-6; ePDF: 978-1-3994-1493-7

2 4 6 8 10 9 7 5 3

Typeset in Adobe Garamond Pro by Deanta Global Publishing Services, Chennai, India
Printed and bound in Great Britain by CPI Group (UK) Ltd, Croydon, CR0 4YY

MIX
Paper | Supporting responsible forestry
FSC FSC® C171272

To find out more about our authors and books visit www.bloomsbury.com and sign up for our newsletters

CONTENTS

	Prologue	ix
1	Preseason	1
2	Australian Grand Prix: Albert Park Circuit, Melbourne	15
3	Chinese Grand Prix: Shanghai International Circuit, Shanghai	31
4	Japanese Grand Prix: Suzuka International Racing Course, Suzuka	44
5	Bahrain Grand Prix: Bahrain International Circuit, Sakhir	54
6	Saudi Arabia Grand Prix: Jeddah Corniche Circuit, Jeddah	62
7	Miami Grand Prix: Miami International Autodrome, Miami Gardens	76
8	Emilia Romagna Grand Prix: Autodromo Enzo e Dino Ferrari, Imola	88
9	Monaco Grand Prix: Circuit de Monaco, Monte Carlo	97
10	Spanish Grand Prix: Circuit de Barcelona-Catalunya, Barcelona	108
11	Canadian Grand Prix: Circuit Gilles Villeneuve, Montreal	115
12	Austrian Grand Prix: Red Bull Ring, Spielberg	124
13	British Grand Prix: Silverstone	133
14	Belgian Grand Prix: Circuit de Spa-Francorchamps, Stavelot	143
15	Hungarian Grand Prix, Hungaroring, Budapest	153
16	Summer Break	162
17	Dutch Grand Prix: Circuit Zandvoort, Zandvoort	166
18	Italian Grand Prix: Autodromo Nazionale Monza, Milan	173
19	Azerbaijan Grand Prix: Baku City Circuit, Baku	181
20	Singapore Grand Prix: Marina Bay Street Circuit, Singapore	189
21	United States Grand Prix: Circuit of the Americas, Austin	197
22	Mexico City Grand Prix: Autódromo Hermanos Rodríguez, Mexico City	205

23	São Paulo Grand Prix: Autódromo Jose Carlos Pace, São Paulo	213
24	Las Vegas Grand Prix: Las Vegas Strip Circuit	223
25	Qatar Grand Prix: Lusail International Circuit, Lusail	232
26	Abu Dhabi Grand Prix: Yas Marina Circuit, Abu Dhabi	240
	Epilogue	247
	Acknowledgements	254
	Index	255

PROLOGUE

It's early Friday afternoon in late May, just before half past one, and smoke-like white clouds are gradually being strung across an increasingly powder-blue sky, releasing the sun to beam down on the city.

Guests weave their way through the leafy public gardens, around the bollards and railings, and the parked mopeds, to make their entrance beneath the imposing pillared and foliage-covered gateway to the Metropole Hotel. The adjacent everyday bank is still open for customers, as are a smattering of decadent restaurants around the corner, though the walkway for now has been narrowed and covered.

Across the road is the Zegg & Cerlati boutique, with its Rolex clock affixed to the wall, and a display window festooned with luxury jewellery and watches. A few metres further up the hill, guests in designer wear mingle on the patio of the Hotel de Paris' Salle Empire, while others enjoy lunch under its canopy. Above them are a plethora of balconies. On some stand spectators who seem just to have left their beds; on others stand spectators in €5000 suits, or more casual wear that probably costs nearly as much. Other balconies remain unattended, as the immaculate frontage, hailing from the Belle Époque, dominates one corner of the pedestrianised square.

Two deep male voices boom out over a Tannoy, welcoming people to the Monaco Grand Prix, first in French, then moving across into English, ready to guide everyone to what lies ahead.

Workers in orange overalls and white helmets, and fire workers dressed in navy overalls with red helmets and an almost golden mirrored visor, survey the scene, perform final checks and limber up for duty. A line of dayglo rucksacks hangs from the mesh fence behind, reminiscent of a school playground; large bottles of water are lined up beneath the shade provided by the branches of a dense overhanging tree; and lunch has long since been packed away. There are fire extinguishers placed upright on the tarmac; a few brooms dotted around – in the event the track needs sweeping of debris and detritus; and an array of colourful flags in a box – yellow to warn of potential danger, blue to warn of a faster car approaching, red in case the session is halted – and all carefully arranged for rapid access.

A kilometre or so away in the pit lane, mechanics make the final checks to their respective cars, fettling the machinery, readying the tyres. The drivers – dressed in stylised and branded fireproof overalls – start to helmet up and contort their bodies beneath the halo, and nestle themselves into the specially moulded seat inside the tight confines of the cockpit.

Around the world viewers will be tuning in, in the early hours of the morning in the United States, or into the night in Australia, or following along on social media.

At 13:30, on the dot, those marshals standing at regular intervals blow their whistles to indicate the race has started. Until 06:00, it has been a public road, and it will be again from around 21:00, but for now it is a racetrack.

Everyone is standing behind an Armco barrier, 90cm (3ft) high, firmly bolted together and slotted into the ground alongside the pavement, above which is diamond-shaped mesh fencing. There are a few gaps, suitable for photography.

The hum of a V6 engine gradually loudens before a flash of red car appears from behind a natural dip in the road, skirting the right-hand corner. It makes a controlled yet quivering slide through the middle of the road as it crests the cambered surface, the back end desperate to kick out sideways towards the barriers, and from the cockpit the driver flicks the steering wheel left to counter the slide. The left of the car brushes mere millimetres from the curves in the Armco, close enough to touch, almost leaving an imprint of the tyre sidewall as a memory. A rush of air hits like a gale, along with a blast of heat, while bits of tyre marble and debris are pinged skywards and spattered around. The machine hurtles by in a flash of illegible branding and a crackle of engine notes, then disappears down the shaded part of the avenue, hidden by the paraphernalia of temporary barriers. It continues towards the hairpin in front of the Fairmont Hotel, where Monaco's railway station once stood, and then into the tunnel, where violent echoes shake the soul of anyone standing within its chamber, before bursting back out into the sunlight adjacent to the harbour, where the tight formation of the white yachts effectively shields most of the water.

Then seconds later another car appears, with slightly different characteristics and a driver with a slightly different level of confidence but flirting with the barriers, pushing as hard as they dare. The next car comes, this one even closer, and everyone takes a precautionary step back away from the barrier before inching a little closer once more to glimpse the next driver on the scene. Then another, and another, and the Monaco Grand Prix weekend is firmly underway.

Some of the rituals have evolved, some of the scenery has changed, the machinery has been transformed and the safety measures have certainly progressed, but fundamentally the end goal for those in the cockpit flinging cars around the same public roads remains now as it was way back in 1929: to drive around the streets of the principality as fast as possible, to beat the competition and win the Monaco Grand Prix.

This is just one corner of one racetrack in a diverse and enthralling championship that now extends across 11 months of the year. Welcome to Formula 1.

1

PRESEASON

'When you start a new season, everyone starts from zero, and it is completely impossible to predict where you are going to be.'
VALTTERI BOTTAS

There was a time when a sparser Formula 1s season began with a race while New Year's celebrations were very fresh in the mind, and when the season concluded just a couple of days away from the year becoming history.

The championship now typically begins in early March, firing the starting gun on a long journey that will incorporate 24 grands prix across a 40-week period. It will hop from country to country, or even continent to continent, in the space of just a few days, the travelling circus normalising the extraordinary shift of people, machinery and other assorted equipment. Finally it will come to a halt in late November or early December, when Christmas trees festoon airport terminals to the sound of festive songs.

But it is not simply a case of personnel reconvening in the paddock in early March and getting on with it. Formula 1 is very rarely in the spotlight during the short, cold and wintry days of January and February – after all, there are no grands prix – but away from prying eyes the wheels of the championship are in motion.

Drivers tend to have the bulk of the Christmas and New Year period off, with minimal if any commitments in early to mid-January, which provides the time to relax and switch off with family and friends. That often means a brief period of inactivity, indulgence and holidaying, before knuckling down at training camps once the New Year rolls into the business part of January.

There are always certain stakes in the ground and expected dates but – save for extraordinary circumstances (yes, this is a nod to the pandemic) – the calendar is finalised the previous summer, setting in stone the destinations and the order of running. Race promoters will always be planning for their next event – and even the one after – before getting a firm date in the diary, but once fixed, they know this is when it will be their turn to welcome the championship. Drivers and teams can begin planning their respective seasons, highlighting whether there are any awkward trips or a quick sequence of races that could place an additional demand on the body, or place the personnel at the factory under greater pressure to produce sufficient spare components. Marketing and media teams will pinpoint grands prix that may be more commercially important or busy due to their respective partners, and also note if and when their drivers have a home grand prix, at which there is always an additional level of interest from local journalists and spectators.

The early weeks of the year are nonetheless a period in which it is important for some to completely switch off.

DRIVER FITNESS

'I would say December is that month you want to do something completely different and not think about it, then slowly process things from the season,' says 10-time race winner Valtteri Bottas. 'Then January, mid-January, there'll be weekly or every two weeks at the factory for sim [simulator] work and meetings, then February is already full gas and we start.'

The winter break is the period in which to get a driver into a better physical and mental shape, assess any setbacks from the previous year, and map out a plan for the upcoming season. Each driver will have individual preferences over training regimes – and the more experienced drivers will have a greater understanding of their own body and its requirements – but maintaining a high level of fitness is merely part of life for most.

That can include elements such as running, cycling, swimming, boxing, tennis, the increasingly popular padel, and usually several hours in the gym, toning and enhancing muscle. Some drivers like winter sports, but the high risk of injury means many have a clause in contracts forbidding such exploits. They must also be conscious to maintain the same weight throughout the year for performance purposes. After a spell where drivers sometimes starved themselves to lose weight, new regulations were introduced that the driver and their seat must weigh at least 80kg (176lb) – which will be 82kg from 2025 – and anything underneath will result in ballast being added to compensate.

'We try and think about things that will become performance-limiting first,' says Jon Malvern, founder of Pioneered Athlete Performance, a sports science organisation that works with Formula 1 drivers, including Lando Norris. 'In the off season it's obviously a very nice time because you have a much clearer block of time to programme and overload, whereas during the year you're everywhere.'

Formula 1 is a sport in which its participants are effectively lying down in a specially moulded bathtub, with their feet on the taps, which is at odds with the exceptionally high level of fitness required in order to compete.

Drivers need to sustain an extremely high level of concentration for as long as two hours, with the potential for high cockpit temperatures – upwards of 60°C (140°F) – in tight confines, while taking a hit from high G-forces, all while making hundreds of split-second decisions and processing information from a race engineer talking in their ear. That level of training – both mental and physical – is essential for success, and cannot be achieved overnight.

'First things first: you actually decide how much time you want to spend on holiday during the off season, and how much time you want to spend training,' says Malvern. 'If you have ten days to two weeks off, generally you will maintain what you've got in terms of cardiovascular fitness, strength, development, muscular endurance. It generally stays where it is. And actually, it's an advantage to taper that down. You'll come back and be stronger or fitter, essentially ready to perform. After that period of time, you start to regress.'

Irrespective of a driver's individual training regime, there will be some aspects of fitness that are mandatory when the season creeps closer. 'Typically everyone needs to go and do neck strength and neck endurance because that's very much performance-limiting and something you don't stress anywhere near as much in day-to-day life,' says Malvern.

Look at a Formula 1 driver and you'll notice the large width of their necks, a requirement in order to cope with the loads placed upon them through high-speed corners. High-tech machines are used to train drivers' necks, while harnesses, precise seating positions and a gradual loading of weight is expertly utilised to prevent strains or injuries.

'You have to withstand 6G, which is equivalent of like 35 kilos [77lb], which is strong,' Malvern says. 'That's the strength element. But you need to be able to repeat it over and over again and accumulate a lot of load. So you have a massive endurance element and that's the harder bit to develop.'

If a race at a high-speed circuit runs without interruption, that could mean 600 or 700 corners at extreme force across a 90-minute period.

Drivers will have regular check-ups – typically during the preseason period – and modern technology allows reams upon reams of data to be collected via various metrics, such as cardiovascular output, strength and muscular endurance markers. This allows training regimes to be tailored to the driver depending on what needs to be tackled, and how each driver prefers to tackle the grind of training.

'Sometimes drivers might only need like four to six weeks because they keep themselves ticking over, and actually you're just trying to get them to accumulate load again so that when they step back in the car, they can tolerate the force,' Malvern says of preseason programmes. 'Some other drivers, especially if they've got a lower training age, you need to allow longer time, and also psychologically they just feel like the more prep time feels better.'

That is an aspect that some drivers simply must learn the hard way.

'Especially the first year I was really struggling physically and that affected the race performance,' says Formula 1 driver Yuki Tsunoda, who now has a two-week training camp in Dubai, a place which is a regular off-season haunt for the grid. 'The more I've been in Formula 1, I recognise how important the off season is; actually, once the season starts it's busy, you can't improve your physical level, it's either maintain or you lose. At the same time, relaxing time is also very important. So it's kind of mixed. But the two-week training camp helps me a lot.'

Each driver will have their own carefully crafted routine to help them get in the best shape possible – and to stay there throughout the year – as starting off on the wrong foot can be costly. Everyone has off-days and makes mistakes, but there's no edit button in a racing car and a driver can't hit rewind on a potentially career-altering setback.

'You go through this long season, you know there's things you need to pick up on,' says Logan Sargeant on the importance of preseason. 'But where's there the time to do it? You can't get into a consistent rhythm, you can't train the way you want to train [in season] because you're constantly recovering, so it's not as straightforward as it may seem.'

Having 24 races for a couple of hours in a 40-week window may not seem overly arduous. But add in three days of on-track driving at the event, extensive media and sponsorship commitments both at the grands prix and away from the track, fan meet-and-greets, preseason testing, the car launch, simulator work, engineering meetings, long journeys travelling on planes – not to mention actually remembering to have a social life away from racing – and it is easy to see where time disappears, and how energy and focus can be gradually drained away as the year develops. Maintaining a training regime – and a level of fitness – is difficult, so to take the next step amid the in-season rigours is even harder.

'It's so taxing to go through the whole season and just learning how to manage yourself over the course of a season is tough,' says Sargeant.

To get ready for that long season, Formula 1 drivers must also find ways of keeping their eye in the game. Unlike most sports, where training with the required equipment can mirror the experience of a match, or help participants improve technique, Formula 1 drivers cannot easily replicate their day jobs.

The development of technology means there are highly sophisticated inbuilt simulators at Formula 1 team factories, where data sets can be inputted to help drivers eradicate weaknesses, while getting a feel for the simulated version of the upcoming year's car. Some drivers also have racing games and low-key simulators at home, allowing them to stay sharp and active. Max Verstappen even travels the world with a gaming rig – aided by the virtual landscape being both a hobby and a business – and fortunately has very strong Wi-Fi connectivity at the luxury hotels in which he stays. There will also be the opportunity to spend some days pinging around in go-karts while some teams have the finances and the ability to arrange private test days in old-specification Formula 1 cars – per the regulations these have to be two or more years old – just to shake off some of the off-season rust.

CAR DESIGN

Each Formula 1 team – officially labelled as a Constructor – must produce their own car for every season. A sport that is fast-moving and ever-evolving (even to the point of exhaustion) depends on a level of technical research that would invoke curiosity at NASA. And the quest to turn the future into the present is relentless hamster wheel of innovation, intuition and intrigue.

The championship has just 10 teams, but they are packed with some of the best engineering brains on the planet, who crave the competition that promises not just success and riches, but also the elusive pursuit of perfection.

In a conventional season a car's concept begins to take shape in the autumn around 16 months prior to the first race, while if there is a new set of technical regulations on the horizon there will be a small working group undertaking research sometimes several years in advance, even if rules can prohibit actual development until

2 January of the preceding year. The FIA – with dialogue from Formula 1, the teams and sometimes power unit manufacturers – produces detailed and prescriptive technical regulations that are almost 200 pages long, covering what feels like anything and everything, so if you do struggle to sleep, then go and check them out. Teams will nonetheless strive to unearth potential loopholes in interpretations and ambiguous wording, often leading to arguments about 'the spirit of the regulations', followed by the issuance of technical directives that act as clarifications before such loopholes are slammed shut for the following season.

Each Formula 1 team is chasing the maximum level of performance, aiming to produce a car that is not just aerodynamically powerful, but one where that downforce can be effectively and consistently harnessed, without sacrificing straight-line speed.

It must have a good balance to provide the driver with confidence, to ensure they can approach a corner and have full faith in the car; there's little point having high levels of downforce if the car is on a knife-edge. The machine must also work in a wide operating window so that it isn't susceptible to minor changes in wind direction or surface temperature, which can be in the low teens (Celsius) during the chilly nights of Las Vegas, and push into the sixties when the black tarmac soaks up the sun's rays in the height of Europe's summer heatwaves. Imagine being a runner and needing to be at top performance regardless of whether it's 5°C (41°F) or 35°C (95°F), and anywhere in between. A car also needs to extract the optimum grip from the softer tyres for short blasts in qualifying while also avoiding degradation across longer race stints, as the car must work in tandem with the four pieces of rubber that are the only thing connecting it with the actual track. It must also work across varying fuel loads, from the handful of litres needed for a qualifying effort to the full tank it carries at the start of a grand prix.

The car must work at a breadth of different circuits where different wing specifications are needed – the skinny wings at high-speed Monza, the 'barn door' wings in twisty Monaco, and everything in between – and not to mention that it must deliver this high level of performance without compromising mechanical reliability. There is also the cost cap, meaning teams have to work within a set budget, and cannot afford to chase down conceptual avenues that may not work. It's a tall order, with modern Formula 1 so competitive that being even a single per cent slower than the fastest car can mean being outside the Top 10.

Teams analyse the airflow of design concepts via sophisticated CFD systems – computational fluid dynamics, which combine simulation and complex mathematics to predict airflow – as well as an allocated number of visits to the wind tunnel, where a 60 per cent scale model is used. This is why correlation is so important. The computers will be delivering thousands of numbers and telling a team that a car will perform in a certain way – and if it gets to the track and does so, then great, the tools are working. If not, and the numbers do not align, and the performance of the car in the real world is lacking due to elements that the virtual tools are not identifying, then there's a whole world of pain and head-scratching coming their way – first to understand the lack of correlation, and then to try and address it. Each team will apply hundreds of sensors to the car to correlate these numbers, but the best sensor is the one wedged in the seat with the steering wheel in their hands.

Chassis shape – effectively the design concept – is usually locked in around September or October of the previous year, after which the chassis has to be homologated, a deadline which has shifted slightly earlier now that Formula 1 has a mandatory nine-day shutdown covering the Christmas and New Year period and prohibiting work from December 24 to January 1 inclusive.

One of the first considerations for each team is the crash test. These are mandatory and they must pass all the safety measures implemented by the FIA. The UK-based teams head to Cranfield, while Italy-based Ferrari and Racing Bulls (RB), along with Switzerland's Sauber, use a facility in Milan. These tests are typically undertaken in December, within weeks of the previous season's final race.

'They get more and more challenging each year,' says RB Technical Director Jody Egginton. 'We talk about rule updates with the FIA so we know what's coming. It's no surprise they are quite challenging. And if you were to fail a homologation test, you've got to react quickly because you can't test the car until it's passed the test. Not even private tests. You've got to have a homologated car to test. So yeah, it's stressful. The FIA have to witness those tests for the nose and the chassis. They have to be there so you can prove you passed.'

Teams book a slot with the FIA and they also have a back-up appointment arranged in case there is a problem during the first test. A delay in getting crash test approval for the required components can have a negative knock-on effect in readying the car.

Once that is passed, further development work can take place. With concepts signed off by the aerodynamics team, the design team can produce a drawing that then goes to the manufacturing department, who create moulds, laminate the piece of carbon fibre, cure it in an autoclave, then machine it into the final shape and send it along to the build team – all in as orderly and efficient manner as possible. It is not quite the same process at each team – some will outsource production of certain components – while the design of the chassis must be undertaken in conjunction with the layout of the power unit, making this process more straightforward for works teams compared to those who buy their power unit.

'For the year we plan to make four chassis,' says Egginton. 'Once the chassis is homologated and crash-tested, there's still an awful lot of work to do on it to turn it into a raceable car. What we send to homologation is known as a Stage One chassis. And then Stage Two is when we bond all the brackets on and all the other bits which are not structural but hold the car together. So one chassis will be in Stage One, the other one will be in Stage Two. And then when that one leaves Stage Two and goes to car build, so we assemble it, the other one goes to Stage Two and the third one goes for moulding to Stage One. So three go through together. Then the fourth one comes a little bit later because that's normally available a little bit later in the year. You need three cars to race: two race cars and a spare chassis, which you can't build up. And the fourth one will come a bit later, mainly because after we've made three chassis, we've got to make loads of car parts to go racing!'

Teams can turn around major components fairly quickly, with one set of the major elements – such as the floor, rear wing, front wing – requiring a week to produce, though teams have to balance resources, logistics and finances.

This hasn't always worked out well for each team. In 2024 Williams was focused on overhauling its design department after decades of old-fashioned practices, and pushed its car build to the latest possible window, but could not manufacture a third chassis in time for the opening rounds. When one chassis was damaged beyond on-site repair in a crash during practice at the third event, in Australia, Williams could field only one car – an embarrassing outcome.

Even as early as preseason, Formula 1 teams will already be thinking several months – or even years – down the line.

'You know already, before the car even runs, and we were in the design process, with the other design managers and our own managers, thinking about, *OK, what do we want to do?* says Egginton. '*How do we want to develop this? Where are we aerodynamically? Which way do we want to go?* So you're already mapping it out and it's already a discussion with the finance people! By the time we go to the first race, we already have updates coming through. Already, as soon as the designers had finished designing the launch-spec floor, they got straight into the new floor. I think, you know, by the first race of the year, I already have a clear view on what we are going to do at race six or race eleven.'

For months the machines are worked on behind closed doors but eventually each Formula 1 team presents its new car to the public. These launches are usually held in early to mid-February, and often lead to an arm wrestle between different internal departments. The technical side are keen to focus as much as possible on manufacturing the car, having little interest in the external attention brought by a launch, while the commercial and marketing departments are eager to maximise publicity for brand awareness and to satisfy partner demands.

LAUNCH

Teams adopt different approaches year on year. McLaren took over Valencia's breathtaking City of Arts and Sciences complex to perform a demonstration run in 2007. Securing a glitzy location for a major unveiling is particularly likely if a team has a fresh partner, driver line-up change or internal reset. By contrast, other teams are happy to keep it very low key, holding the launch at their factories or even sometimes merely posting a computer render of their new package online. Some years, a couple of teams have unveiled their new car in the pit lane an hour or so ahead of the start of the preseason test. The approach can vary year on year. In 2023, Ferrari held a grand launch at its test track at Maranello, with fans and media present, unveiling the new car and then immediately demonstrating it by completing a couple of laps. In 2024, the new car was presented in a 90-second video on social media.

Car launches in the modern era are increasingly season launches, often revealing just the car livery. The teams are keen to keep technical secrets, designs and bodywork under wraps for as long as possible, so it helps that they maintain control of the imagery. Desperate to avoid leaks, teams ensure that unveilings are co-ordinated down to the minute. Each team's two cars must be run in the same livery – though in-season special designs are permitted – and for fans this is among the most intriguing and anticipated moments of the entire year. It is also the moment where the employees of each team can see the car that will represent their hopes and dreams for the upcoming campaign. There is always renewed optimism and freshness when nothing has yet been written for the year ahead.

'It's always funny when you're about to go to the final race and a friend is on the phone and they're like *Oh three months of holiday . . . no it's not!*' says Will Ponissi, Head of Communications for Sauber Motorsport.

'The moment the racing starts or the car goes out in testing in Bahrain, you're in execution mode until Abu Dhabi, there's not a moment to breathe or sit down and plan. So you're not just planning the month in that two months before, some of the plans and projects you have for the whole year, you're planning them there – there's a lot more going on than just the launch.

'It's your story – *who do you want to be for a year?* All that stuff takes place in the planning period. *Which markets are we particularly targeting? What is important for our partners?* There's usually a lot of partners being announced, you want them announced before you display the car with all the logos on them. We need to talk and align with partners, get things approved, there's a lot going on. You want to give as much notice as possible to media, as you want them to have everything you need, as you know who you can work with in terms of confidentiality information.'

Launch season can be a scattergun time of year, with attention quickly switching to the next team in the spotlight. All 10 teams tend to present their new seasons within a short time frame and sometimes over half the grid launch in the same working week. Launch complete, the emphasis swiftly turns to on-track running.

TESTING

Formula 1 used to permit unlimited and unregulated testing, but in a bid to curb unnecessary spending amid spiralling costs, regulations were introduced in the late 2000s. These were gradually refined to the extent that there is just a sole preseason test, lasting three days, at which teams are allowed to field only one car. The teams are also allowed two filming days – in order to capture rights-free footage for their own content distribution away from Formula 1's strict broadcast arrangements – at which a maximum of 200km (124 miles) of running is allowed. These filming days are now utilised by teams as a form of car shakedown. They are often arranged before the start of testing, with the focus on ensuring the car runs smoothly, to avoid any potentially damaging setbacks at the official test. It is an exciting moment for each team as a long-term project bursts into life for the first time.

'We're trying to do things in a very quick amount of time, but you sort of get used to that intense period, and there's a lot of people heavily involved in the process,' says Jody Egginton, Technical Director for RB. 'Altogether in the last days it all just meets in the build shop. So the car itself gets built quite quickly in the end. It's based on just-in-time principles. Nobody wants to sit there with a finished car for a week because then you say, *Why didn't we use that time in the wind tunnel!*'

After the shakedown it is off to preseason testing. This now typically takes place in Bahrain. Spain's Circuit de Barcelona-Catalunya was more convenient for scurrying new parts from the factories – usually located in the UK and Europe – but Bahrain now has the advantage. The state offers better facilities, the ability to run longer days in representative conditions and, perhaps most importantly, more stable weather in February – it is never going to snow in Bahrain, as it has done in Barcelona, and Formula 1 cars do not exactly function well in icy weather.

Days at testing for mechanics are lengthy – this is not a normal 9–5 job – and teams tend to have a rotational day crew and a night crew, because there is no curfew as there is at grand prix events.

'It's super important,' says Egginton about testing. 'It's the first day back at school for drivers and engineers and mechanics. The things you want to learn about the package are massive. You have to be organised and use every minute.'

Engineers will have a lengthy run plan to digest, while this is the only opportunity for drivers to understand a car and its nuances ahead of the first event. Teams typically split running equally, meaning each driver has a maximum of 12 hours of track time, which usually equates to only about 200 laps. This is one of the only sports in the world where training time in current equipment is so restricted.

'You really have to make the most of it,' says McLaren driver Oscar Piastri. 'I think, in some ways, because we have the three practice sessions per weekend and we have so many races, you kind of have an opportunity to learn on the fly. But if you're in a championship position, for example, you don't really have the time to do that at the start of the season. It's very, very important to get that testing right. And also for the team, just making sure everything works. It's a brand-new car for everybody at the start of the season.'

'Testing is a great moment to learn all you can about your package: you start to learn the ways its set-up works, but it's only at the first race that you kind of get some kind of a picture,' says Valtteri Bottas. 'When you start a new season, everyone starts from zero, and it is completely impossible to predict where you are going to be: it's the beauty of sport, especially nowadays, with the fine margins currently separating the teams.'

Nevertheless this a period of smokescreens, body language analysis and team personnel asking: 'What do you think of —'s car?' because this is a first opportunity to glance at the solutions adopted by rivals. Behind the wheel, the drivers will already have an inkling as to whether or not they have the package they want. For several months now, they may have known – thanks to the simulator numbers and meetings at the factory – that things weren't going well. And this can sometimes be the moment when reality finally hits.

Now the teams will have reams of GPS data and analysis to run through thousands of computer models and systems. Hopes will be either rising, or sinking, as the first grand prix of the season looms on the horizon. As Valtteri Bottas puts it, 'When you start a new season, everyone starts from zero, and it is completely impossible to predict where you are going to be.'

Pre-season testing, held here in Bahrain, is a chance to see the new cars

2

AUSTRALIAN GRAND PRIX: ALBERT PARK CIRCUIT, MELBOURNE

'I grew up about 10 minutes from Albert Park...I used to watch them on TV, then run out to the backyard and hear them going round.'

OSCAR PIASTRI

The longest awayday on Formula 1's calendar takes the travelling circus to Australia and to Victoria's popular capital, Melbourne.

For personnel it means a 24-hour trip each way, via one of the Middle East connecting hubs that begins to feel like home across the opening rounds. Even once the plane reaches Australia there are still another four hours to go, and an enormous 11-hour time zone shift that plays havoc with the brain, but fortunately Melbourne is a buoyant city that reinvigorates any weary traveller. It is a vibrant city, compact and easy to navigate, renowned for its hearty breakfasts and coffee culture, and thanks to Melbourne's relatively late timetable – to make the schedule as palatable as possible for the European audience – team members can easily take advantage. The most popular haunts require a lengthy queue, particularly on weekends. The line for the pastries at the renowned Lune bakery – founded by former Formula 1 aerodynamicist Kate Reid – always stretches around the corner of the street (and its outlet in the fan zone in Albert Park quickly runs out in the mornings!), while in the evenings the restaurants on Southbank adjacent to the Yarra River are packed to the rafters (and you can't avoid a traditional chicken Parma). Others prefer to hang out just to the south of Albert Park, in the beachside commune of St Kilda, famous for its miles of golden sand and colourful beach huts, and a backdrop of Melbourne's cityscape, where promotional activity with local drivers often takes place.

Melbourne held Formula 1's opening round from 1996 to 2019 (aside from 2006 and 2010) and it still feels like the real start of a fresh year even in seasons when it is no longer the opener. Under its contract – and due to the impact of Ramadan, the dates of which shift each year, in both Bahrain and Saudi Arabia – Melbourne will host the opening round again on at least four occasions between 2025 and 2037 – no less than such a city and an event deserves.

Australia has produced two World Champions in Sir Jack Brabham and Alan Jones, both of whom have bronze busts at Albert Park. Melburnians rally behind the local driver, even if local can sometimes mean a racer from thousands of miles away. In recent years Australia has had Mark Webber, from New South Wales, and Daniel Ricciardo, who grew up in Western Australia's Perth.

Ricciardo first visited Melbourne's grand prix in 2002, as a 12-year-old already participating in kart races, and attended the race at which debutant Webber finished a shock fifth for backmarker team Minardi. He was enthralled by the spectacle.

'I remember walking around the general admission, just seeing the cars and trying to watch,' Ricciardo said in an interview in 2024. 'And I remember at that age I was very obviously in awe. But also I did look up to these guys as, like, superhuman. I was like, *I can't do this*. I was like, *These guys are freaks!* And I was fascinated.

'My dad's a fan of the sport as well. So we would try to do as many grands prix as we could. I would say we would go probably every second year. Back in those days when it was the V10 engines, you'd arrive at your hotel in the city and you would hear when the cars were on track. And I was like, I just remember as a kid being like, *Dad, we're wasting time. Let's get to the track!* I was a full "keeno". I was all about it. Yeah, I was a bit of a nerd!'

Ricciardo scored his maiden points in Melbourne, in 2012, and two years later claimed a first career podium at the race – though was disqualified afterwards for a technical infringement.

'That was like . . . that race for me –' he says – 'whether I got the trophy or not . . . Yeah, as far as I was concerned, up until the disqualification, I raced well, I qualified well, I stood on the podium. Obviously I was upset. But I just proved so much to myself on that weekend where I was like, *f—k!* I was like, *Yeah, I can . . . I can do this*.'

Dealing with a home grand prix can nonetheless be a challenge for a driver owing to the heightened interest and attention.

'From the outside everyone's like, *Well, Daniel, you know you've got the podium, so can you win it now* and all this,' Ricciardo explains. 'And we always unfortunately seemed to start the season quite slow and never really be a podium car. And then as the sport grew and as I guess my place in the sport probably grew, the more attention was on and the more that, you know, you get pulled left, right and centre, more sponsors and more this and that. And yeah, so there was definitely a period where I didn't enjoy it as much as I wanted to. And that was unfortunate because home is a special place, but I think it took a little bit of going through the s—t to then kind of just figure out how to manage a home race.'

Thanks to his engaging personality Ricciardo will always be a popular figure, especially in Melbourne, but the city now has a bona fide home boy in the shape of McLaren Racing's Oscar Piastri. Piastri won titles in Formula 3 and Formula 2, before stepping up to Formula 1 in 2023, claiming two podium finishes as a rookie. He was born and raised in the affluent suburb of Brighton, a mere 10km (6 miles) south along the coast from Albert Park, and became the first Melburnian to race at Melbourne's grand prix in 2023.

'I grew up about 10 minutes from Albert Park, so especially when there was the V10 and V8 days, I used to be able to hear them from my house in the backyard,' Piastri says. 'I used to watch them on TV, then run out to the backyard and hear them going round, so it was cool.'

Despite Piastri's proximity to Albert Park, and his enthusiasm for motor sport, his first time at Melbourne's grand prix wasn't actually until he was a teenager. Under a Formula 1 initiative, the local sporting body selected promising young drivers or kart racers to be 'grid kids' at grands prix – and it was Piastri's first foray into Formula 1.

'I went in 2015: I was a grid kid for Daniil Kvyat,' says Piastri. 'And he broke down on the formation lap, so I wasn't a very good luck charm for him! It was very cool. Going on the grid was an amazing experience. I think at the time I didn't really understand how privileged I was to be able to go on to the grid. Of course, now I get to do it every weekend, but it's a slightly different form! It was really, really special. You know, just seeing the F1 drivers up close and I think especially at that time the way the sport was, and also being so young, I kind of just knew everyone as the helmets in the cars and not really as the people.'

A young Piastri also got to indulge in his fandom elsewhere at Albert Park, watching trackside at a left-right kink, which drivers approach at 325km/h (202mph) and where spectators can watch from mere metres away.

'I got to see how quick the cars were,' he recounts. 'That was a pretty special moment to watch. I remember I went out to the back chicane at Melbourne, the fast one, watching them go through there, and we go through there even quicker now. It was a pretty special moment.'

Australia's isolated location in the racing scene means those who do want to make it to Formula 1 have to be fully dedicated to the cause. The likes of Webber, Ricciardo and Piastri all relocated to Europe in order to pursue the dream.

'It was tough at times,' Piastri says. 'And, you know, of course you miss your friends and family back in Australia. But at the same time, I knew that that was what I had to do to try and get to Formula 1 or, you know, even to become a professional racing driver. I knew my chances of becoming a racing driver were much higher in Europe. So I kind of knew what I was sacrificing for and that made things easier.'

It is also a country – and city – which is among the favourites of the non-Australian drivers.

'I think it's the lifestyle, the culture, it's pretty easy-going,' says 10-time grand prix winner Valtteri Bottas, who triumphed in Melbourne in 2019. 'The people, I feel like having fun is very high on their priority list in life, which is good. It's a beautiful place, Melbourne and its surroundings, it's a beautiful country – and it's big, there's so much more to see. It's a good vibe also and at the right time of the year the weather is usually pretty guaranteed.'

Australia initially joined Formula 1's calendar in 1985 at a street circuit around South Australia's Adelaide. It was placed as the final round of the campaign, cultivating an end-of-school aura, and it produced its fair share of enthralling title showdowns. That was most notable in 1986, when a dramatic puncture denied Nigel Mansell a shot at the crown, and again in 1994, when Michael Schumacher and Damon Hill collided, delivering the first of seven world titles for Schumacher.

Neighbouring state Victoria swooped and from 1996 Formula 1's Australian round moved to Melbourne, with a street-style course navigating its way through leafy Albert Park. There was also a seasonal shift as Melbourne was placed as the opening round, meaning Australia went straight from holding the finale in 1995 to being the curtain-raiser in 1996, won by Damon Hill after debutant Williams teammate Jacques Villeneuve ceded the lead due to an oil leak. It has regularly been the season-opening grand prix across the near three decades of its history. Michael Schumacher, the most successful driver, took four wins in a five-year spell through to the early 2000s, while the last five editions have been won by five different drivers. Melbourne's Turn 1 has also been the scene of several pile-ups over the years, most notably in 2002, when eight drivers were eliminated in a large accident. A tight bottleneck of a medium-speed corner, with grass, gravel and walls close by, tackled by drivers a little rusty after the off-season, can be a recipe for collisions.

Melbourne had faced fierce rivalry from adjacent New South Wales, which frequently makes noises about trying to hijack the event for Sydney – the level of scheming in this championship sometimes makes politics look like child's play – but a long-term deal signed in 2022 secured Melbourne the grand prix until 2037, warding off the threat from other states.

That Melbourne was able to claim an iron grip on Australia's grand prix was a remarkable post-pandemic turnaround.

Formula 1's 2020 season hung on the precipice as the pandemic developed through the second week of March. The circus flew out to Melbourne – which was scheduled as the first race in 2020 – and on the previous Sunday the women's T20 final took place in front of 85,000 people at the Melbourne Cricket Ground. But by Wednesday distancing was being enforced at media events and on Thursday the talk was about whether the weekend would go ahead as planned. Discussions between a multitude of parties – drivers, team bosses, team owners, promoters, government officials – sparked into life late in the evening when McLaren withdrew after a member of its personnel returned a positive test. Meetings were held at the Crown Hotel through the night as various proposals were tabled even as information and the associated advice was evolving. Some of the proposed solutions were madcap, including running one practice session featuring the only two teams firmly up for competing. Friday morning dawned and team personnel arrived early, some determined to continue as normal until told otherwise, others beginning to pack up. One team continued practising pit stops as per their schedule, even though two drivers had already taken off from Melbourne's Tullamarine Airport to head home, and fans gathered at the main gates, which remained closed. As rain began to fall, an outdoor press conference attended by senior figures from Formula 1 and the Australian Grand Prix finally confirmed what had been expected for several hours: cancellation. This was the first time in 35 years that a grand prix had been abandoned while on location; the last time was Belgium in 1985, when a recently resurfaced track broke up. Travel arrangements were now hastily made as the majority of the paddock sought to return to Europe before the situation deteriorated.

Australia's strict international travel restrictions meant the 2020 event could not be rescheduled, and 2021's round was postponed and then called off because Formula 1 was not willing to undertake the mandatory quarantine period upon entry. The Australian

Grand Prix Corporation (AGPC) had to remain alert in order to secure the event's future.

'We had to control the controllables, we had to make sure we were controlling the uncontrollable, as if we didn't take control others might have set the path and the destiny of the grand prix and it might have been a different outcome,' says Andrew Westacott, CEO of the AGPC from 2011 to 2023. 'It was a tough couple of years. There was the fear of not coming back to Melbourne if we did not host the 2022 event.'

Fortunately, 25 months after everyone traipsed away from Albert Park on a gloomy and windswept Friday lunchtime, Formula 1 rocked back into sunny Melbourne in April 2022.

'International reputations are earned over a very long period of time in the events industry and Formula 1 has a lot of choices to host its events, and it needs events to make its model work from a media rights perspective and a promoter fees perspective,' Westacott says. 'I can't recall the actual wording or details, but there was an expectation from the AGPC and F1 and we had to make sure that was understood by the health and sports authorities, and the cabinet and the Premier [of Victoria], that we had to deliver the event. In 2022 we had 419,000 spectators and we actually signified the opening up of Melbourne, we got our mojo back: it was a significant event, the sun shone, record crowds, and we pushed well into 125,000 people on race day. It was a bookend: we were the first event cancelled globally, and certainly the most significant cancelled in Melbourne at the start of Covid, and that in 2022 was an opening up, and it gave the events industry a lot of confidence that you could host an event.'

Usually, more than 400,000 spectators attend Albert Park's busy four-day weekend and there are a plethora of activities and support categories. In recent years the likes of Formula 2 and Formula 3 have travelled to Melbourne, joining up with Australia's renowned and more gruff Supercars series, and other local-level sportscars and single-seater categories. It means, in a rarity for a race weekend, that track activity starts on a Thursday morning, in order to squeeze in such a busy on-track schedule. Organisers also used to hold quirky activations, such as a relay race between a Formula 1 car, a V8 Supercar and a sportscar, while the demonstration run of a V10-powered two-seater Formula 1 car is typically the 7 a.m. alarm call for those not up already due to jet lag, the high-pitched screams from the engine crackling through Melbourne's sky. The officials also set up Melbourne Walk, a narrow bendy strip of tarmac between the car park and the paddock entrance, which is shaded by the overhanging eucalyptus trees, and which has fans crammed along one side to greet the drivers at dawn and dusk.

'I was thinking about the Oscars, the red carpet, the mixed zone, where the paparazzi are close to the actors, fans see them, they make a spectacle about it,' Westacott says. 'The drop-off for the paddock was right outside of the turnstiles, like in many locations, so the fans would be there morning and night, but they'd only get the smallest glimpse. We moved the drop-off point 150 metres [490 feet] further back and rather than the red carpet we created the Melbourne Walk through the gumtrees of Albert Park. It's become quite a spectacle now and something we're very proud of.'

It enhances the atmosphere even further as huge cheers greet the drivers – and increasingly the team principals, too – each morning, with the most hardcore of fans arriving early to grab prime position along Melbourne Walk.

Formula 1 slots in as a major event in a city that energetically embraces large sporting spectacles.

Melbourne has a Sports and Entertainment Precinct that is just a short walk or tram ride from the CBD (central business district). It houses the 100,000-seater Melbourne Cricket Ground, home to four of the 18 teams from the Australian Football League (AFL), with fixtures often overlapping with the Australian Grand Prix weekend, allowing fans and personnel alike to soak up a fixture, try to understand the rules and try to stave off the jet lag. Melbourne Park is home to the Australian Open, one of only four tennis Grand Slams, which takes place in January.

Piastri, a budding cricketer in his youth and a supporter of AFL team Richmond, agrees that sport makes up a huge part of the city, with Formula 1 a key component of Melbourne's calendar.

'Being from Melbourne, I always thought about the grand prix, I would always watch the Aussie Rules grand final, I will watch the cricket,' he says. 'I didn't really like tennis that much growing up, or other events, but you'd always kind of just watch them because they were on and they were in town, that kind of thing. So I think we always get behind our big sporting events, and I guess there's kind of an event for everyone in Melbourne, because we've got pretty much one of everything in a major sport. We're a sports-mad city, we're the sports capital of Australia, definitely. Which I know will upset a lot of people that aren't from Melbourne! But, you know, we have everything. And I guess that's part of what you can do in Melbourne as well. You can see an Aussie Rules game, you can see a cricket match, you know, you can see the tennis if it's on. There's a lot of sports to see and I think we're a very active country in general. But Melbourne I think really epitomises that.'

Formula 1 takes over a public environment for the week – and longer due to the set-up and pack-down time involved – and there is a constant need for justification in terms of using public funds.

'There was a never-ending and incessant challenge to make sure people knew and understood how big F1 is on the global stage, and therefore how big it is for Melbourne's major events calendar, and I think recent years and the extension of the contract has shown that,' Westacott says. 'There was a lot of hard work for a long period of time to make sure we were appealing. We're an event that's actually been imposed on Melburnians: the Melbourne Cup, the city has grown up with it; they've grown up with the Australian Open, the AFL, with cricket; F1 is imposed on us, paid for by the taxpayer, so it has to be seen to produce benefit. One thing we've done is show Melburnians how important this event is to the international reputation and how it does contribute economically from a branding and marketing point of view, and also a cultural and civic pride point of view.'

There are still some locals who bristle at Formula 1 being at Albert Park, but organisers work with residents to minimise disruption. Planning for a grand prix can take the whole year, but the actual build process begins around the time of the New Year. It's approximately a 12-week process, dismantling the temporary infrastructure taking around half that period.

'We only close the park on the Monday of event week. The Melbourne public are very accommodating and we have to make sure that support is never taken for granted,' says Westacott.

'In the early years Albert Park wasn't as splendid a park, back in 1996, as it is now. On the back of the AGP the government invested money into facilities and infrastructure and grounds; it's a much more beautiful environment than it was back then. People didn't know what to make of Formula 1 because they didn't really realise the importance and standard of it. Now generationally we're seeing that the Formula 1 AGP has a place and in fact there's only two big international events Melbourne has – the AGP and the tennis. Yes, we have other events, but they don't project internationally like grand slam tennis and Formula 1.'

Melbourne took over from the very popular Adelaide, wavered a bit when it relinquished the honour of the first race, and faced an extremely real threat to its long-term future during the pandemic. But it now has one of the highest race attendances, is an adored venue on the schedule, and has one of the longest contracts on the calendar.

'We get judged by attendees against permanent infrastructure, so we have to be up there with the best of the permanent infrastructure but in a temporary environment,' says Westacott. 'The other thing it does is it allows the government and the AGPC to make investments that will have a payback via increased reputation and tourism. But it shows this event projects Melbourne on an international stage; there's not many big events you can get for a city. Every couple of years, Adelaide comes up, Sydney comes up, but the best place to host them from a crowd and location point of view is Melbourne, so 2037 was a very important commitment from the government.'

It is, as the locals will quip, a great place for a race.

The location of Formula 1's opening round has changed across its 75-year history.

The likes of South Africa, Argentina, and Brazil have kick-started a new season over the years.

Since the mid-1990s, with the exception of the pandemic-hit 2020 season, Formula 1 has either begun its campaign in Australia or Bahrain.

Round 1 is the sole event of the season at which – even with reams of data, preseason knowledge and internal expectations – there is still that slither of hope that *anything* is feasibly possible.

By this point, testing has provided a hint of who is quick, and who is not, with the increased access through TV feeds, reams of GPS data and the more revealing body language already giving indications of the likely pecking order at the start of the season. Teams still strive to hide as much as possible to keep their powder dry for as long as possible.

Formula 1 grands prix are four-day weekends: there are media activities on the first day (usually Thursday), followed by practice (Friday), qualifying (Saturday) and the race itself (Sunday). The tension builds throughout the early phase of the weekend, with greater attention than usual placed upon the hot laps and the longer run simulations during practice at the opening round, when all 20 cars are on track together for the first time. That anticipation gets ramped up on Saturday, with everyone excited for the opening qualifying session of the year.

This is where the big reveal takes place. It may only be one session, at one type of circuit, in a certain type of condition, but it is the first time in the season where everyone has the same objective, with just enough fuel for a hot lap in the car, and the softest tyres being run. *Who's got it? Who hasn't?*

'I mean, the real acid test is the first qualifying session of the year,' says Jody Egginton, Technical Director for RB. 'You know, people can form a view, and we form a view of where we are in testing, and what other people's fuel loads might be, and what they're doing and what they're not doing. We all make analysis of competitors, but it's not until you get to that first qualifying and everyone takes the fuel out and they're going for it, then you see really where you are.'

No one can hide after the first qualifying session of the year. Yes, someone may have made a mistake and lost a few tenths. Someone may have had a reliability glitch. And maybe the season-opening circuit does not suit the car's hoped-for strengths. But the pure underlying performance of the car – and the driver – is there for all to see, as the stopwatch never lies. This isn't the time to form a definitive judgement – there is still nearly a year to go – but for the most part drivers will now know where they're likely to be competing throughout the season. Those at the front have had their preseason optimism validated, those further down the grid have had their worst fears confirmed, and technical bosses of those teams will already be feeling the pressure to remedy the situation.

'The most nerve-racking point for me tends to be qualifying for the first race, because that's when you really find out where you stand – it's not something you can tell from winter testing,' says James Key, who has held technical directorship roles with McLaren and Sauber. 'You never know where your competition are [in the winter], because they've done exactly the same job as you, they have the same ambitions as you have, they want to move forward and progress.'

Wherever it is, the first race day of the season is among the most anticipated, with nervous energy and excitement fizzing through the paddock. After all, everyone starts with zero points, and there are reputations, honour and millions in prize money on the line. The approximate 2000+ personnel who swipe through the paddock turnstiles will all have aspirations for the year ahead, none more so than the 20 drivers who call the cockpit their office.

'When you haven't raced for a few months, it feels a bit more intense,' says Daniel Ricciardo. 'You can spend time in the simulator driving the track, but you can't practise racing intensely, wheel to wheel with 19 of the best guys in the world, and that makes for a pretty exciting feeling at the first race of the year.'

Three months have passed since the final chapter was closed in the previous campaign and it is the first litmus test of all the lessons learned across the past year, the past winter, in the long and dark days at the factory, hours in a simulator in a windowless room, and the sometimes tedious preseason running.

It is the first race day walk into the paddock in new uniform, the first pre-race engineering and strategy meeting, the first pre-race meal, the first drivers' parade, the first pre-race warm-up. This is what drivers have been thinking about during long and lonely months in the gym in the winter; it's the date on the calendar which designers will have ringed when pondering concepts on a computer screen midway through the previous year; it's the date championship organisers, and broadcasters, will have looked towards when the schedule was unveiled the previous summer. Everything in theory remains possible – if only for a few more hours.

The tension builds in the paddock through the day, a nervous sense of excitement buzzes through the hospitality suites, while VIPs and guests stake their claim for the prime vantage points, as the spectators gradually fill the grandstands, serenaded – or rather deafened – by the music being blared out by a DJ.

With 40 minutes to go, the music halts, the engines are fired up in the garages, and the field of 20 rolls out to complete reconnaissance laps of the track. The drivers ensure the sophisticated systems are operating as expected and get a feel for the track conditions before cruising up to the rear of the grid. They switch off the engines and almost ghost towards the assembled mechanics, who wait like expectant parents trying to pick out their child at the school gates. Each of the 20 cars gets jacked up on to a wheeled 'tea tray' by around a dozen mechanics, and the components susceptible to overheating are equipped with myriad cooling devices straight out of a sci-fi film, then pushed to their assigned grid spot, the group of eight or so mechanics moving like an army to strike a path through the crowd. Unsuspecting ditherers – irrespective of status – will simply be swatted out of the way like pesky flies. The greater attention is always on the front-running teams, who carry an aura of superiority, and the backmarkers are sometimes overlooked, wary that they are likely in for a tough day at the office. The car effectively becomes a hospital patient, hooked up to machines and laptops festooned with data charts and graphs to check temperatures, pressures and levels, monitored by the engineers – the car's doctors – who undertake final checks under the watchful eye of scrutineers, while the temperature-sensitive tyres are wrapped up in their protective blankets. Drivers, now out of the cockpit and striving for a slither of quiet sanctuary next to the car, update race engineers about the lie of the land – any last-minute information could glean vital tenths of a second in-race – while performance coaches ready ice packs and water bottles, and undertake final checks of the driver's helmet and visor.

TV presenters from around the world – chased by a camera operator weighed down by bulky equipment – jostle for position, eager to snatch a word with a driver, team boss – shadowed by a press officer – or an often mystified celebrity, while dozens of tabard-clad photographers capture their final grid photos before scurrying away to the location from where they will shoot the first lap. The 20 drivers dart their way to the front of the grid, to stand alongside dignitaries and representatives, for the national anthem, sometimes accompanied by a flypast.

The grid clears of non-essential personnel, leaving only the mechanics, engineers and drivers, who complete final talks, fist bumps and rituals before the drivers clamber beneath the halo and into the cockpit – an arena of solitude and tranquillity after the controlled chaos of the previous half hour.

As the branded clock at the end of the pit lane strikes the hour, the 20 cars head off on the formation lap – a single lap of the track at reduced speed – leaving behind an army of mechanics, who scuttle back to the respective garages to prepare spare parts, check wheel guns and lay out fresh tyres in case first-lap misdemeanours require an unscheduled pit stop. Two minutes later the cars slowly return, one by one, completing final burnouts and system checks to maximise performance off the line. Everyone gathers around the closest TV screen, all watching the same F1-produced feed, while in small boxes above the main grandstand the voices of commentators quicken and louden as they raise the anticipation for those back at home across all four corners of the world.

The 20 cars form up, 10 on either side, each machine separated by 8 metres (about 26 ft) – and when the final car is in position, a marshal at the back waves the green flag, indicating to the starter that the procedure can begin. On goes one red light, the engine notes rise, then comes the second red light, third, fourth, until all five lights are illuminated. *Pause. Hold breath.*

Almost two hours later, press officers and performance coaches wait for their respective drivers behind the FIA's garage at the top of the paddock. There is a quiet hubbub to the landscape, with polite chit-chat, before the press officer escorts and accompanies their driver to the TV pen – even as the driver is trying to cool down, take a moment, sipping on a drinks bottle to recover lost fluid because no amount of winter training can prepare for the real thing. Not unlike a display at the zoo, drivers make their way around the inside of the horseshoe-shaped arena, the TV crews and interviewers on the outside, each with their own allocated spot in eight separate zones, beneath the bright heat-emitting lights. Straight out of the cars, and with barely any time to think, they will convey their thoughts to the various interviewers – sometimes repeating the same lines eight or more times – as everyone gets their soundbites. And like a zoo, the drivers come afterwards to the more congested written media zone, with journalists trooping across from the refrigerated media centre where the race was watched, digested and dissected. Maybe for someone fifth place was a total disaster while elsewhere 11th for another driver was a standout achievement. The top three drivers celebrate in the pit lane, and then on the podium, on the other side of the paddock building. There is the slightly muffled sound of the national anthems – the winning driver's followed by the victorious constructor's – before the trophy presentation and the spraying of champagne – or non-alcoholic rose water in some territories – accompanied to the energetic beats of the overture from Bizet's *Carmen*.

Post-race those who have exceeded expectations cut a buoyant mood though caution against arrogance – 'It's only one race, we can't take anything for granted' – all while inwardly knowing they are set for a superb season, in which trophies will be matched by an enhanced reputation. Those who have endured a disappointing round either strike the positives, clutching at straws amid faint hope, or concede that the year ahead will be long, torturous and potentially embarrassing. The right words at pivotal moments not only highlight a driver's mood, but will also affect the factory-based members of each team, who will be in varying states of emotion depending on the result. The social media posts that inevitably follow invariably contain more uplifting and motivational slogans.

For those in the paddock the reality of the season ahead has truly kicked in. Personnel know they are in for either a year of collecting trophies, and the probability of a financial bonus for Christmas, or a season of struggle, where keeping up motivation will be one of the main challenges. Live TV broadcasts eventually end, the media mingle around for team boss interviews that feature varying degrees of optimism, and guests gradually drift out of the paddock turnstiles. Personnel don fluorescent jackets as the pack-up process swiftly accelerates – having actually begun during the race – leaving in the air just the beeps of reversing forklifts and various objects clanging against the ground, being thrown into piles, and then chucked into pallets for the next event.

Eventually the key personnel will filter out of the paddock, whether to head back to the hotel, or straight to the airport to fly home, ruminating on the varying fortunes experienced a few hours earlier. Some will ponder if the trophy cabinet at the factory will need expanding, while others will wonder if they'll even make it into double-digit points all season. Round 1 complete, just the 23 to go.

Melbourne's skyline sits over St. Kilda's waterfront,
a stone's throw from Albert Park

3

CHINESE GRAND PRIX: SHANGHAI INTERNATIONAL CIRCUIT, SHANGHAI

'Despite Netflix being technically unavailable in China, some of Gen Z got hooked on Drive to Survive.*'*
FRANKIE MAO, FORMULA 1 JOURNALIST

China was supposed to be a new frontier for Formula 1 in the early 2000s, but even as the sport enters its third decade since its debut here, the country remains something of an untapped market.

Formula 1 explored China in the late 1990s, recognising the country's potential emergence as a global player, but a proposed race at the Zhuhai International Circuit never fully got off the ground. The circuit was constructed, but the governing body, the FIA, deemed that the venue was not up to the required standard to host international events on Formula 1's level.

Instead, a new project, in the developing Jiading district on the outskirts of Shanghai, was launched. That came soon after capital Beijing had won the bid to host the 2008 Summer Olympics.

'Formula 1 wasn't really known to the majority until the late 1990s,' says Frankie Mao, China's sole full-time Formula 1 journalist. 'At the time Shanghai, known as the economic capital of China, had successfully promoted the city by hosting international sporting events, most notably football and tennis, but nothing would match the ambition of the city compared to having a Formula 1 grand prix.'

Formula 1 arrived in Shanghai for the first time in 2004 and was greeted by an enormous venue, with a vast paddock and pits structure that had two winglike platforms traversing the circuit at either end of the pits straight, while team buildings were arranged along tree-lined paths and alleyways designed to mimic ancient Yu Gardens. The layout of the circuit drew inspiration from the Chinese character shang, which roughly translates into English as 'ascend'.

China was initially installed as a late-season grand prix – and even hosted the non-title deciding final round in 2005, when the championship had already been won by Fernando Alonso – and it has delivered its fair share of championship-defining moments. Michael Schumacher won his 91st and final grand prix in 2006 at Shanghai, and one year later Lewis Hamilton's hopes of becoming a rookie World Champion collapsed when he slid helplessly into the tiny gravel trap at pit lane entry in a rain-hit race. He made amends a year later with a controlled win against rival Felipe Massa, putting him on course for his maiden title.

Formula 1's championship-winning teams of the past decade and a half both claimed first victories at Shanghai: Sebastian Vettel gave Red Bull Racing its first win in torrential conditions in 2009 and three years later Nico Rosberg dominated to restore the modern iteration of Mercedes to the top step of the podium.

Vettel's victory in 2009 was the first of China's revised calendar position, after the event shifted from a late-season affair to an early encounter, where it has remained. China was an annual event until 2019, with the round planned for 2020 the first to fall by the wayside once the seriousness of the pandemic was realised. China's strict Covid Zero policy, and the effective long-term closure of its borders, meant the event was cancelled in 2021, 2022 and 2023.

Formula 1 finally returned to China in April 2024, after an absence of five years, and when it did it marked the first time that local fans had one of their own on the grid.

China was not represented in Formula 1 until 2012, when the little-known Ma Qinghua was parachuted into a test driver role for teams at the back-of-the-grid HRT and then Caterham, but he dropped from the scene once his limitations behind the wheel were realised.

Hope subsequently rested on Zhou Guanyu. He had left Shanghai as a youngster and relocated to Sheffield, South Yorkshire, in order to compete in the renowned karting categories, his ambition being to reach Formula 1.

'It's unrealistic that if you're racing at home you're going to get to your dream in Formula 1,' Zhou says. 'You start racing there, then you have to go to Europe and race with all the best young junior drivers and make your way through.'

Zhou was part of Ferrari's Driver Academy in Formula 3, moving across to Alpine's young driver scheme in Formula 2, and eventually making his Formula 1 race debut in 2022 with the Alfa Romeo-branded Sauber team.

'I went to the first-ever Chinese GP in 2004, I really enjoyed it,' says Zhou Guanyu. 'The first two years it was only the very first starting point back home, so not many people knew about it, the tickets were being given away, the popularity wasn't that high. We started in general as a country many years far behind compared to Europe, but it was the very beginning. In recent years it has been building massively, and 2019 was a big step because it was the time when we had the 1,000th Formula 1 grand prix, at the track, and a ticket was really hard to get. Then unfortunately it stopped for a few years because of the pandemic, but now there's a lot of people into racing and Formula 1.'

Zhou's presence in the sport had an uplifting effect in a country that the F1 authorities understood to be lucrative, but had yet to fully understand.

'The reasons behind it were various,' says Mao on Formula 1's slow growth in China. 'Long story short: it was a result of the mismatch between two giants that had been misunderstanding each other from the beginning. It started to change in 2017 when Liberty Media took over, and in 2019 the 1,000th grand prix took place in Shanghai, as Zhou did a demonstration run in Xin Tian Di and it was the first time an F1 car ran on a public road in China during the race weekend. Prior to the race on Sunday, Zhou did three parade laps, and when he climbed out of the car and stood on the cockpit holding a national flag, enormous cheers erupted from the grandstand.'

The landscape changed through the pandemic, and at the same time Formula 1's mental shift, recognising that it had not fully understood the Chinese landscape, facilitated a different relationship.

'Despite Netflix being technically unavailable in China, some of Gen Z got hooked on [the documentary] *Drive to Survive*,' Mao explains. 'In a way it was lucky for both Zhou and F1 to make . . . history at the right time. Back in the day, the casual audience would barely have had a clue why he wasn't able to win or be running in the front. Now, the new generation of fans, having learned enough from *Drive to Survive*, appreciate how it works to succeed in F1.'

At Zhou's debut home event, in 2024, he finished only 14th, but was greeted with a hero's welcome, and Formula 1 even created a space for him to stop post-race on the pit straight. He climbed out of the car and was overcome with emotion, crouching on the tarmac.

'As much as I say I want to focus and treat this like a normal race weekend, but when you see the crowd from Friday, all packed into the grandstands and every time you come out of the garage they are all cheering you on the grid,' Zhou said. 'This just gets me, you know, this journey to get here – I'm just super proud and honoured to be finally the first Chinese driver to compete in the Chinese Grand Prix for 20 years. Emotional of course, it was a very special moment for me.'

China is, after all, home to a fifth of the world's population, who increasingly have disposable income. It is a country which has grown as a global power, having held both the Summer and Winter Olympics.

'F1 eventually cracked the US market, which has three races on the calendar now, which led some teams to shift their focus to the States, from the sponsorship perspective,' Mao says. 'However, F1 does understand the China market more than before . . . F1 and teams have learned that if you want to approach a Chinese company for potential partnership, you have to be present in China – at the least on China's social media sites. So, including F1, half of the teams on the grid have their own Chinese social channels because none of them want to miss any opportunities in China.'

But the situation is complicated. Formula 1 and China are still getting reacquainted with each other, on account of the five-year break, and the future remains unclear. The boom in the United States, and the financial powerhouses emerging in the Gulf, have altered the landscape, and China is no longer the prospective gateway to new markets – and therefore money – that it seemed to be at the turn of the millennium. What's more, increasing political suspicion of China, as well as the Great Firewall of China, means communication – and therefore accessing the Internet – can be a barrier to smooth working conditions.

'The current Chinese GP contract will run out in 2025,' says Mao. 'The host fee is never a problem as long as the event continues to deliver what the city wants to see. Having said that, if one day the Chinese GP stops, will I be surprised? No, I won't. [In] the end, it's a business, so it has to benefit everyone involved. Though, personally, I'd love to see the Chinese GP stay as many years as possible, as it's become one of the icons for this city.'

F1 is carried by China Central Television – the state broadcaster fittingly known as CCTV – and though broadcasting is essential for any sport, surviving is not enough. On television a sport must also thrive and Formula 1 is no different.

Across the decades Formula 1's coverage has gone from grainy highlights of races that last just a few minutes to access all areas, all the time, everywhere, by everyone.

Broadcasters pay Formula 1 a fee for the rights to show the sport in their respective country or territory. Some of these deals are more lucrative than others depending on the level of competition, the access desired by the broadcaster or the territory in question. They can then produce their own off-track content, and have innovated certain features and analysis across the decades, but all broadcasters – roughly 70 of them – fundamentally receive the same screen product: Formula 1's world feed.

Formula 1 produces the world feed, which is operational at the circuit from Thursday to Sunday night and acts as the eyes on the circuit's activity. Once you see the official introduction five minutes before a track session – accompanied by the theme music and drivers greeting you in various states of seriousness – that'll be the trigger for every broadcaster to switch from their own product into the world feed, which is watched live at every race by around 70 million people.

Formula 1's production team has two key locations throughout a grand prix weekend. The first is at the circuit, with its 25 × 15 m (82 × 49 ft) windowless canvas Event Technical Centre located in the paddock. Formula 1 has two of these tents that leapfrog each other at grands prix, such is the timescale involved in setting them up, though all the internals – including the dozens of TV screens that make it resemble NASA HQ – are taken to each event. The second facility is its permanent Media and Technology Centre at Biggin Hill, in south-east London, the former airbase utilised by the championship for decades. The ETC on location cuts the track mix from the trackside cameras, which is the bedrock of the world feed and needs to be done quickly, while the team's responsibilities at Biggin Hill include elements such as team radio, cutting the replays and managing the on-screen graphics. The two units work in tandem to ensure there is a constant feed through the weekend, while there are reserve teams and back-up procedures in places in the event of a total failure.

It is a huge undertaking. Over 20 times a year a stretch of circuit between 4 and 7 km (2.5–4 miles) long – some of the tracks temporary and city-based, others across challenging terrain – must be rigged up such that a live 4K feed can be transmitted to hundreds of territories.

'If you think about a football match you have two teams, one ball, on a pitch,' says Formula 1's Director of Broadcast and Media Dean Locke. 'We've got 10 teams, 20 balls in effect, and on a really large pitch . . . there's lots of things happening simultaneously, at a very fast speed, over a big real estate.'

Formula 1 has around 50 cameras, which includes those affixed trackside, cameras installed inside kerbs, cameras in the pit lane and paddock, and helicopter shots. There are also 90 on-board cameras (approximately four or five on each car, the most prominent being the T-cam, which provides a forward-facing view just above the cockpit), of which 22 are able to be broadcast live.

A director controls what is transmitted live on the world feed, but there is a large team at the ETC and at Biggin Hill scrutinising various angles, on-boards and live timing, to highlight whether any incident needs to be swiftly cut to or clipped up for a replay. Formula 1 also has access to team radio communications, the best of which are picked up, the aim being to broadcast them within a lap to ensure it remains fresh and relevant for the narrative.

'The director tells them what they want, the person there finds that on the map, tells his track director we want him on camera 9, they then cover that car. Cutting track cameras is an art,' says Locke. 'The main director here has that as a source, he will say: *I'm taking track mix*. There's stuff happening not live on the camera, so if, for example, Hamilton brushes the wall at Turn 9, they'll flag that, and then we'll decide to put a replay on it.'

There are emergency systems in place to ensure the world feed remains active in case of a technical failure. The threat of a typhoon in Japan in 2019 meant the entire ETC had to be relocated inside one of the pit garages in a matter of hours on Friday evening, from where the broadcast of Sunday's race took place without a hitch. One year in Germany a drainage issue meant sewage was swilling through while work continued, while one broadcast in 2021 continued smoothly for the outside world despite two dozen people being taken out of action by Covid.

The F1-controlled world feed has been in place since 2004, and over the following years Formula 1 has gradually taken over the production of a larger number of grands prix from national broadcasters. By 2011, when Japan's Fuji Television transferred its coverage to Formula 1, all bar one event was sorted. Monaco remained an outlier and as the years passed the inferior quality of its product – frustrating direction, outdated angles, missed action and incorrect narratives – was plain for all to see. One year there was an in-weekend battle for control of the buttons such was the frustration from Formula 1 over Télé Monte Carlo's direction. Under terms of a new contract, from 2023, Télé Monte Carlo relinquished duties to Formula 1.

Locke explains the old broadcast model. 'Host countries did the "world feed". Germany was RTL, for the UK it would have been ITV or BBC. It was a really good model because for that local broadcaster it was a blue riband event . . . they put a lot of onus and effort into that. But by the early 2000s we were seeing sports broadcasting develop in a lot of ways: Sky is turning up, Fox and ESPN in America, and really developing sports broadcasting. The level of what the viewer wanted was developing quite fast. You don't want any local bias . . . some local bias was there . . . and even though they might get the best sports director in the country in, they're probably only doing one, maybe two events, a year. We felt, *How could we improve the broadcast?*

There had been prior experiments. Formula 1 launched F1 Digital+ in 1996 to produce its own enhanced coverage targeted at pay-per-view broadcasters, but it was a commercial failure in an environment not quite ready for such a product and closed in 2002. But that provided lessons on how coverage could be enhanced.

'We had some techniques and we were pretty well-versed at it,' Locke explains. 'We were also moving to territories where they didn't have a local broadcaster well-versed in international sporting events. We thought, *Well, why don't we put together a full-time team that looks at the race every week?*'

The product has developed across the past two decades — which has seen the advent, boom and evolution of various social media platforms — while in 2018 Formula 1 also launched its own bespoke product, which is available in certain territories, called F1TV.

'Over the race weekend we're operational eight hours a day, so keeping ahead of that content, and keeping it to the level it should be, is pretty demanding,' Locke explains. 'When I first started we had 16 races, the majority in Europe, now we're at a 24-race season with 50 per cent flyaways.'

The championship continues to develop its on-screen product, with meetings and feedback after every grand prix, as it evaluates and assesses new initiatives. These may relate to a new angle, different camera techniques or innovative graphics.

It is a laudable achievement, capturing machines moving at over 320km/h (200mph) across a landscape that sometimes stretches out across several kilometres in varying terrains, and delivering that to viewers across the world in just seconds. It can be frustrating that some angles focus too much on sponsorship hoardings – for obvious financial purposes – but static trackside cameras truly convey the breathtaking speed of Formula 1 cars. And the relentless quest to innovate, and not rest on laurels, is admirable.

'We're always moving,' Locke says. 'My bosses are ex-F1 people, so they're used to developing at a fast rate and they expect that from us in a media scape.'

The live broadcast of a full qualifying session, or indeed a grand prix, is a huge undertaking for a team that is determined to deliver the best product.

'I remember when I was directing the world feed, I still found the pressure of that and I was never blasé about it when I turned up on a Sunday,' Locke says. 'The directors that do it now are still equally like that.

'Editorially we're after action, and the higher up the order, the better – if there's a battle for second and third, we'd probably prioritise that over seventh and eighth – but we're not going to show second and third if they're not battling. There's great stories in F1.'

That also extends to catering for the audience, and ensuring that a breadth of knowledge and interest levels are considered.

'We don't want to patronise someone who understands a lot about Formula 1, but we don't want to alienate people with a graphic that's too difficult to understand. I know a bit about sports broadcasting yet I'd watch some sports and go, *I'm not sure what that graphic is about*,' Locke says.

While the world feed provides the visuals of the on-track action, it is then up to the job of a commentator to help sculpt the narrative. They spend the weekend in cosy booths trackside, littered with microphones, headphones, a commentary unit – essentially a control panel of buttons and faders – and all the assorted cables.

'Practice sessions are very much like a podcast,' says Alex Jacques, Formula 1 commentator for F1TV and Channel 4. 'It's far more of an intimate relationship with the audience there, it's for devotees – the ones who know everything inside out, you're explaining a lot less, maybe recapping a lot less, and then it's more expositional as the weekend goes forward. Qualifying is pure *say what you see*; you're just talking about sector times, talking about improvements, talking about head-to-head records, and talking about the significance to the driver on that day in that moment.'

The biggest moment of any Formula 1 grand prix event – which most people will be watching – is race day, most specifically when the five red lights go out.

'At the race you almost reload again and start from scratch, because you're getting the majority of the audience watching that weekend for the very first time, it's your broadest audience,' Jacques says. 'But the hardest thing on a normal weekend is lap one. It is free-form, they are right together, 20 cars, and describing, and choosing what to describe, your economy of words, your driver recognition, it is a challenge every half second for about 45 seconds until you get to the part of the track where everything settles down. Thereafter it's not as punchy as qualifying, there's not a headline every few seconds, you've got 90 minutes on average to talk about a whole range of things. Sometimes you won't see the leader, so you need your narrative and research, and the ability to talk to everyone – from the 10-year-old fan who's just discovered the sport for the first time to the grandparent who's left it on after watching something else, or a brand-new fan who has switched on after watching *Drive to Survive* and has thought: *You know what, I'm going to give this F1 a go and understand it.*'

Getting an insight into the drivers' mindsets, and their reactions, is also crucial for broadcasters. Drivers have a routine of mandatory interview duties after practice sessions, qualifying and the race. It can be a tricky and fast-paced environment in which to operate, with drivers filtering through ad hoc, all with varying emotions depending on how their qualifying session or race has unfolded.

'It's something I've learned over years, to gauge their mood,' says Lawrence Barretto, a journalist and reporter for F1TV and its official website. 'You also get to know them separately and build a relationship separately. You can see their tells, their facial expressions, and even if they've had a bad day they will understand the questions you're asking are fair questions and you're not trying to dig them into a hole – you're offering them the opportunity to respond.

'You'll have drivers come up to you and say they've had a retirement, but it wasn't their fault, so they're annoyed, but you can tell they want to talk about it, so you have to think quickly on your feet and phrase the question. You might have seen them walk past already with their helmet on, so you can see they're not in the mood to talk at the TV pen, so you might phrase the question a bit softer to get them talking, then you can ask a little bit more of a pointed question – so it's making loads of quick decisions based on your experience of that person, and what they're like, and also knowing you've got to get a good answer out because you're supplying content to the viewers – they want to know what the story is.'

While Formula 1 is fundamentally highlighted by its on-track product, there remain off-track elements for the world feed team to consider and to emphasise.

'The great thing about F1 is, you put on the TV, or watch it on a device, see a clip, and you know where you are,' Locke explains.

'Each race is very different. At Silverstone there's a huge festival spirit and it's our job to replicate that to the fan: *Ah I want to go to Silverstone, as it looks really cool for a festival spirit*, but then we know when we come to Singapore it's a completely different vibe and we've got to show that. A traditional race like Spa has its own kick; the big push for us is to replicate what the race is and what the race is about; wherever we are, we have to show that event. In Miami they did driver launches, it's a completely different vibe, that's Miami, that's what that race is.

'We had a shot of David Beckham on a pit stop. Now he's seen some stuff, we had him look at a pit stop and go *Wow*, because they are phenomenal. We took Keanu Reeves out trackside and when he sees those cars under braking, he's not acting, he's blown away. They're cool moments.'

Broadcasting a high-speed sport means there are inevitably worrying moments after accidents in which a driver's condition is unknown. Protocols are in place after accidents to ensure that a situation is treated cautiously and with sensitivity. Live TV broadcasts in high definition were not so prominent during the era in which serious injuries and fatalities were unfortunately more commonplace, but over the decades there has been a gradual recognition not to show replays, or potentially uncomfortable scenes, until the wider picture is known.

'A lot of our camera operators do motor sport regularly, and sometimes they're our first message – they will automatically react, they'll open a wide shot, they'll say it's bad and we'll cut away,' Locke explains. 'Us and the producers here will start to engage with the FIA, on giving us an update, as we'll be holding back on replays. If we think it's a risk, we'll wait and clarify with the medical team before we cut back to it.

'A good example is Romain Grosjean's accident in Bahrain. Firstly, we didn't expect to see Grosjean jumping out of a ball of flames! We didn't run the replay straight away, even though we saw him, as firstly he was limping, and secondly, *Is he OK, because maybe he's about to collapse on the floor*, so we were still checking with the medical team. The other part to that was, *OK, Grosjean is fine, but what about marshals, photographers, cameramen?* So we had to do our housekeeping to make sure everyone is OK. We're pretty well rehearsed, the broadcasters know why we're doing it, we have a [communication] line that means we have someone telling them what we're doing, that gives them a rough idea if we have an extreme situation.'

It can also be a tough situation for the person tasked with covering a serious accident.

'If you get a really long red flag and it lasts for two hours and everyone's alright, it's the easiest thing in the world, because your boss cannot come to you the next day and go, *That's a load of nonsense*. Of course it's a load of nonsense, nothing was going on, you were filling dead air,' says Jacques. 'The hardest thing is when track activity is suspended and you know someone is hurt, and information suddenly gets sparse. In that moment it goes from being a sportscast to a newscast, and the rules of the game change. Your element of speculation disappears; you want to be reassuring, you can't tell people it's going to be OK – it becomes very intimate with the audience because you're having to directly relay information in real time and be immaculate with your broadcasting. You can't make any mistakes because you could drop your team in it if you say the wrong thing.'

Formula 1 produces hours of live coverage for three (or four) days, 24 times a year, in diverse locations, ensuring that viewers across the world never miss a beat, and are able to see anything and everything just seconds after it has happened. It is, perhaps, the hardest major international sport to broadcast live and it is a critical component that underpins the entire championship.

China joined Formula 1's calendar in 2004, and Shanghai has hosted every event

4

JAPANESE GRAND PRIX: SUZUKA INTERNATIONAL RACING COURSE, SUZUKA

> *'It is such a thrill. The undulations and the flow to this circuit, it gives you such a great rhythm when you're racing. So it's always a joy to come here.'*
>
> GEORGE RUSSELL

Japan is an island nation where space is at a premium, let alone space for parking – and Kei cars come into their own. Practical and compact (no wider than 1.48 m, no longer than 3.4 m) they remain top sellers, though rarely exported. Their manufacturers are all global brands – from Toyota and Mitsubishi to Suzuki and Subaru – and Japan's automotive industry is one of the world's largest.

When it comes to Formula 1, though, there is only one brand that stands out: Honda. In a sport where the local spectators blend unbridled enthusiasm with a unique respect shown towards the protagonists, Honda is the only Japanese brand to have enjoyed success. It has had a roller-coaster journey, with multiple departures and comebacks, through regulation changes or financial constraints, or merely at the whim of the senior management at corporate level.

Honda has had two stints as a works factory team, in the 1960s and late 2000s, but by itself it never built on glimpses of promise.

Jenson Button won the 2006 Hungarian Grand Prix, suggesting Honda was on an upwards trajectory, but poor cars in 2007 and 2008 stunted its prospects. Behind the scenes the global economic downturn of 2008 impacted major manufacturers and in December Honda pulled the plug on its Formula 1 team. Nonetheless, it still planted the seeds of what blossomed into the title-winning Brawn Grand Prix in 2009, which was acquired by Mercedes-AMG for the 2010 season.

Honda's most prolific spells have instead come as an engine manufacturer, to Williams in the mid-1980s, McLaren in the late 1980s and early 1990s, and to Red Bull Racing in the early 2020s, when it became a key component of dominant title-winning sprees. It has also had lower ebbs, most notably during its second McLaren stint in the mid to late 2010s, when its products married the twin disasters of chronic underperformance and desperate unreliability. Its critics pounced and Honda's famous slogan, *The Power of Dreams*, was unkindly but accurately reversed when yet another engine expired.

Honda has had its own team, supplied teams and supported Japanese drivers financially both to reach and race in Formula 1. And its involvement has gone even further. In the late 1950s Honda founder Sochiro Honda visited one of the company's plants in the town of Suzuka and reckoned that to improve the performance and safety of his products a bespoke test facility was required. Locations throughout Japan were scouted before the selection of a site in Suzuka, a choice that was sweetened by a pledge of support from the municipality. Honda deliberately chose a hillier area to develop, to avoid destroying any flat land suitable for agriculture, especially for growing crops. A circuit with a crossover section – a rarity in motor sport – was constructed and opened in 1962. Honda went on to evolve the circuit into a complex featuring an amusement park, a hotel and a traffic education centre for Japanese citizens, all crammed into the tiny patch of land that the corporation owns.

Yet when Formula 1 did visit Japan, in 1976, it did so not at Suzuka but at Fuji Speedway, in the shadow of the famous mountain. This proved to be an iconic race in Formula 1 history, because it was the first race broadcast live on TV, due to the interest in the title battle between James Hunt and Niki Lauda. It paved the way for the championship's boom via the medium across the following decades. The rivalry between the chalk and cheese James Hunt and Niki Lauda now played out on TV sets around the world. Lauda, just weeks after his fiery near-fatal accident at Germany's Nürburgring, withdrew due to torrential weather conditions. Mario Andretti won the race, but James Hunt overcame a puncture in the closing stages of the race to secure the third place that he needed to be crowned World Champion.

Fuji hosted just one more race, in 1977, which was won by Hunt, before Formula 1 embarked on a prolonged absence from Japan.

Eventually, Formula 1 found its way to Suzuka in 1987 and – bar an unpopular two-year return to Fuji Speedway in the mid-2000s when Toyota stumped up some cash, and an even more unpopular two-year absence in 2020/21 due to the pandemic – it has remained at the circuit ever since. In fact, Suzuka has been the venue for some fierce and contentious title deciders, having been placed either at or towards the end of the campaign during its initial years on the calendar. Ayrton Senna and Alain Prost twice controversially and spectacularly collided at Suzuka, settling titles in 1989 and 1990, while other legends of the sport such as Mika Häkkinen, Michael Schumacher, Sebastian Vettel and Max Verstappen have all sealed world championships at the venue. A generation of European fans have fond memories of rising in the early hours of an autumnal Sunday morning to switch on the TV to watch the season finale.

As Formula 1's calendar evolved and expanded, Suzuka edged further away from the end of season. In 2024 it moved from its traditional autumnal slot to a chillier spring berth, as part of the championship's long-term strategy to better regionalise its schedule, in a bid to be more sustainable, aiding the logistics and reducing the burden on personnel. That meant Suzuka coincidentally aligned with Japan's famous sakura season, when the cherry trees blossom in vibrant pink and white, adding a gorgeous backdrop to some of Japan's major landmarks. Any major calendar relocation of an established grand prix nevertheless takes some getting used to, as if Christmas were suddenly plonked down in March, or Easter in November. Personnel have come to associate races with a certain time of year and Suzuka, so long placed at the denouement of a campaign, is now one of the season's formative rounds.

That move has also lessened the chance of severe weather impacting the event. There are days when Suzuka has been baked in sunshine, and other days where the facility has been shrouded in a gloomy, grey and rainy mist, and teams have to be on their toes to react to the weather, optimising whatever conditions come their way. There have even been typhoons. On three occasions in the last couple of decades Suzuka's timetable has had to be reshuffled due to the threat of a typhoon. Most recently, Saturday activity in 2019 was cancelled in its entirety and instead the drivers, ever the competitive bunch, organised a FIFA tournament at their hotel. Permanent signs in the nearby coastal commune of Shiroko – also printed in English and Portuguese – warn of the risks of earthquakes and tsunamis, and to seek higher terrain if the ground does shake. Suzuka, thanks to Honda, is at least on a hill.

For the Formula 1 community, Japan is always among the favourites. During the build-up, the bulk of the paddock even squeeze in a day or two in Tokyo – or other cities such as Kyoto or Osaka if there is a greater period of time off.

'My Japanese experience is always amazing. I love being out here, I love being in Tokyo, it's one of my favourite places to come to because the culture is just so special,' says Lewis Hamilton. 'It's such a deep and rich culture here, the people are amazing, the food is fantastic, the cities are just all unique and offer such different history. So there's a lot more to see here than what we often get to do on our short journeys.'

Take yourself to Tokyo to wander around Shibuya – famous for its Crossing – or Shinjuku – renowned for its colourful lights and buzzy 24/7 lifestyle – and it is commonplace to bump into other paddock personnel and drivers, checking out the vast array of activities the city has to offer, before making the journey to the relatively rural Suzuka, located in the Mie prefecture. That means understanding and adapting to Japan's excellent but sometimes baffling railway network, where stations sprawl across several levels and have a labyrinthine complex of avenues and alleyways, shopping centres and stores, servicing multiple companies and lines. Everyone, though, loves the sleek Shinkansen bullet trains that dart through the Japanese countryside at 320km/h (200mph), before a quick final hop to Shiroko, the closest town to Suzuka. There is an immediate sense of Formula 1's importance to the district: outside the station is a *Welcome to Suzuka* sign in the style of the black-and-white chequered flag, while stone plinths are engraved with the handprints of the previous podium finishers and the respective results across the years. The Japanese Grand Prix is by far the biggest event for the region in a country where baseball, football (increasingly so) and sumo wrestling are the most popular sports.

There is a heightened sense of camaraderie at grands prix such as Japan, since personnel frequently stay together in sleepy districts such as Yokkaichi, Shiroko or Suzuka itself, creating an almost youth hostel-like vibe. These small towns are a world away from the energetic and colourful metropolises for which Japan is better known and this is a *like it or lump it* sort of weekend: hotels are few and far between, the rooms often cramped and uncomfortable – reflecting Japan's preoccupation with miniaturising everything to ensure no space is wasted – and while modern technology bridges the gap between Japan and the rest of the world, pointing, gestures and Google Translate is still needed, particularly when trying to make sense of symbols and imagery on hand-written menus in cosy restaurants in non-tourist areas.

Food is an enormous part of Japanese life – and the Formula 1 experience – and while sushi, ramen and Kobe beef are world-renowned, there are a plethora of dishes and regional recipes.

'Everyone knows Kobe beef, right,' says driver Yuki Tsunoda, who is as passionate about food as he is racing. 'Wagyu is not just Kobe beef. This Kobe beef is famous around the world, but at the same time, for example at Suzuka, there is a beef brand of Wagyu called Matsusaka beef. And each place, each prefecture, there's a brand of Wagyu beef. Normally the name is coming from the place, that prefectural name or whatever, the town name. Yeah, Kobe's the town name. And the Matsusaka is a part of the town name of the Mie Prefecture. So each brand of beef will taste a little bit different. There's types of fat, the sweetness of fat, also how much fat is inside in general. It's a little bit different for each brand. In Japan, we all know this. When you go to Suzuka, don't eat Kobe, eat Matsusaka beef. Because Matsusaka is actually coming from that town. So it's much fresher. And actually, for me, it's better. If you go to Kobe, which is like slightly farther away from the Suzuka, and closer to Osaka, eat Kobe beef, but some of most of the prefectures, there's a brand of Wagyu. It's good to taste!'

Don't argue with a culinary connoisseur.

Suzuka's popularity is aided by its ribbon of tarmac being relished by every driver. It is the only figure-of-eight circuit on Formula 1's calendar, meaning its layout resembles a Scalextric fantasy track.

'Every lap is pure joy in Suzuka,' says Fernando Alonso. 'The nature of the track, the high-speed first sector, these kinds of cars come alive thanks to the downforce and the grip we have.'

Suzuka is full of high-speed corners, rapid changes of direction, and elevation changes packed into its congested, narrow and leafy 5 kilometres (5 miles), with limited grass and gravel run-off adding to the challenge for drivers.

'Suzuka is definitely one of the very greats of all time,' says George Russell. 'It is such a thrill. The undulations and the flow to this circuit, it gives you such a great rhythm when you're racing. So it's always a joy to come here.'

This is a track where a driver needs total confidence and trust in the performance and balance of the car in order to attack with full commitment, particularly through the rapid changes of direction in the first sector, where drivers rarely drop below fourth gear.

'This track is absolutely amazing – the feeling you get in qualifying through the first sector is amazing,' says Charles Leclerc. 'If you get out of the [correct operating] window of the car it becomes a lot more challenging and might be not as fun! But it's always something special to be driving Formula 1 cars around Suzuka.'

Accidents at Suzuka are rarely small, though the fearsome 130R kink – slightly sanitised since the turn of the century – has witnessed colossal shunts.

'The room for error is pretty small,' says Lando Norris. 'There's a lot of consequences as soon as you go off pretty much anywhere. It's not just tarmac [run-off] everywhere, they've still got grass, you've got that bit of risk, that element of risk, which I think is always genuine; it always feels very natural and definitely adds to, say, a qualifying lap, with how much risk you want to take. You're always a bit afraid of running too wide, things like that, because you know if you put a wheel half a metre too wide, it's game over. So it's exciting. It definitely brings a few more nerves while putting a lap together and driving.'

Suzuka's quaintness is evident when traversing its winding trackside paths to reach some of the best viewing points.

Due to the way the layout winds its way through the natural topography, as well as the necessary safety barriers, fences and marshal posts, some sections of trackside spots for photographers are navigable only by leaping over those barriers – though fortunately into a section of circuit where cars would never reach – or by twisting around some of those fences. That leads on to Spider Alley, an uneven narrow strip between the crash barrier and safety fence which is covered in branches, twigs and Japanese knotweed. The nickname arises from the thin black, yellow and red Jorō spiders that inhabit the area – not poisonous, but not exactly friendly.

While Suzuka is an adored circuit, it played host to one of the most tragic events in recent Formula 1 history, when rising star Jules Bianchi suffered a serious accident in wet conditions in 2014. Bianchi's car left the circuit and collided with a recovery vehicle that was tending to a previously crashed car. Bianchi suffered severe head injuries in the impact and never regained consciousness. He died in hospital, in his native France, nine months later, becoming Formula 1's first driver fatality in two decades. Each year flowers are left behind the base of the barriers that he struck, along with a simple note: *JB17, Never Forgotten.*

'It's a very special place and whenever I get here I have, somewhere in my mind, Jules,' says Charles Leclerc. The Leclerc and Bianchis are close family friends, and Leclerc – born eight years after Bianchi – is his godson. 'I think about Jules very often because he's been the person who helped me to get there back in 2010. He had spoken to

Nicolas, my manager, in order for me to be supported to get to F1. And he's been the game changer in my career. Before that we have been extremely close and both of our families are still very close.'

Formula 1 fans all over the world are passionate, but that attachment reaches a new level of devotion in Japan. 'I remember my first event at Suzuka and being completely blown away by the volume of support from the fans,' says Pierre Gasly, who raced in Japan's Super Formula championship in 2017, before stepping up to Formula 1. 'It's incredible, so special and warming, and every year we are there for Formula 1 it is the same.'

Spectators arrive dressed in elaborately creative clothing, including homemade merchandise and headgear, and it is commonplace to witness 'drivers' wandering around the fan zone. In past years spectators have come dressed in full replica race gear and helmets in tribute to the likes of Max Verstappen, Lewis Hamilton, Ayrton Senna, Sebastian Vettel and even some less heralded Japanese racers from the 1990s, and they are spotlessly detailed. There are throngs of home-made signs, messages and banners while some fans provide gifts to teams and drivers, including to the less heralded members of the paddock. Teams regularly receive sweets and chocolate with team members' faces printed on modified replica packets of KitKats, Pringles and M&Ms.

The spectators are always unfailingly polite, respectful of those around them and hugely enthusiastic about every detail and nuance of the sport, flocking to the circuit from the dawn of Thursday morning until after the sun has set on Sunday evening. Spectators line up alongside Suzuka's access road each morning and night, waving and bowing to anyone and everyone who enters and exits the circuit, irrespective of status. They even remain in the grandstands into the night – clutching mobile phones and glow sticks – to watch the mechanics at work, determined to pay respect to everyone who contributes to the inner workings of the sport. During on-stage interviews in the fan zone, spectators sit cross-legged on the floor, much like a school assembly, and listen intently to each speaker without interruption. It is impossible not to be energised and warmed by the genial atmosphere fostered through the weekend.

'The people are so friendly,' says Lewis Hamilton. 'They're really just a different energy of crowd, they're so . . . you get the sense they're very close to their inner child. They have all these cool little toys or cars on their caps, and they go all out with their little styling. It's really great!'

Every time any driver leaves the pit lane in any session, there is a polite ripple of applause around the circuit and, while local representatives are adored, no non-native driver or team feels the support here is partisan – certainly not in comparison to some events.

'Even if you don't feel them when you are driving, you feel the atmosphere in the weekend [is] different here, about how enthusiastic they are, how they live the full weekend,' says Fernando Alonso.

Suzuka also goes the extra mile in terms of creativity for the official products it sells in its shops. You can buy cushions shaped as a giant spanner, stick-on tattoos and Suzuka branded underwear. You can eat sweets in the shape of gravel, crunchy 'asphalt' cookies, and dried squid mangled to resemble discarded tyre rubber. In a landscape where sports merchandise can be exploitative and overpriced, such products are both charming and memorable.

Suzuka also organises for local schoolchildren from different primary schools to be affiliated with a Formula 1 team. They visit the pit lane and garage on a Thursday and often leave drawings and messages of support that teams pin up for the remainder of the weekend.

'They don't just turn up, they have lessons at school where they learn about the sport and the team,' says Ayao Komatsu, Haas Team Principal. 'They come with flags and personalised messages and they take time to learn what we're doing. They also learn English to be prepared to speak a few words at track and for many it's the first time they meet non-Japanese, European people. They just want to say that one sentence and interact, it's amazing. I think Japanese people as a nation, when they see something new, they want to know more about it. I think in their culture if you're exposed to something, they go really deep.'

Japan has never had a World Champion – or even a grand prix winner – but there is always passionate support for its own representatives, from the likes of Aguri Suzuki to Takuma Sato, and more recent locals such as Kamui Kobayashi and Yuki Tsunoda.

Hopefully, the passion and devotion to Formula 1 in Japan, and the presence and commitment of Honda, means Suzuka – which is contracted through 2029 – will be a much-loved fixture of the championship for many, many years to come.

The 'Welcome to Suzuka' sign outside Shiroko railway station

The main thoroughfare of Tokyo's popular Shinjuku district

5

BAHRAIN GRAND PRIX: BAHRAIN INTERNATIONAL CIRCUIT, SAKHIR

'When Bernie said: "OK, why am I coming to Bahrain?", the story is, there was a map. And he was told: You are the world championship right now. You're flying to Japan, Australia, and you're flying over this part of the world where up and coming this will be a player. And we should have a race on the map here. And Bernie said, "That's the middle of the water." Yeah, there is an island there on the map!'

SHEIKH SALMAN

When Formula 1 broke new ground in an emerging market it did so in one of the world's smallest countries, which had little prominence in the international sporting landscape, but which had grand ambitions.

Bahrain is an island nestled between Saudi Arabia and Qatar in the Persian Gulf. At just 780km² (300sq mi), making it only slightly larger than the Isle of Man, and with a population of only 1.6 million, it is dwarfed by its more influential and powerful neighbours, particularly close ally Saudi Arabia, to whom it is connected via a 25km/16 mile-long man-made causeway. Manama, in the north-east of the nation, is the bustling capital, a financial hub that is often visited at weekends by wealthy partying Saudis, who cannot access alcohol within their own borders and so impose themselves on Manama. Districts in which personnel stay, such as Juffair and Seef, are a mixture of high-rise developments and barren rubble wastelands, and outside of Manama the bulk of Bahrain is largely flat, featureless desert. Most of the southern half of the island is uninhabited, bar the occasional seafront luxury hotel resort – where those who wish to can remain within the same complex all weekend – as well as ugly oil fields and off-limits military bases. The Tree of Life, a 400-year-old arboreal mystery, is one of the main tourist attractions, continuing to grow and blossom in spite of its desert surroundings and despite no obvious source of water. Very occasionally the whistling winds blow a sandstorm Bahrain's way, choking the air and shrouding the scenery in a beige haze.

Bahrain pulled off a coup in 2002 by securing a long-term contract with the championship, becoming the Gulf's first Formula 1 event in 2004. That journey towards Formula 1 was assisted by a three-time World Champion, who had a chance meeting on a Concorde flight with Bahraini royalty in 1999.

'The idea started with His Royal Highness, the Crown Prince,' explains Sheikh Salman Al-Khalifa, Chief Executive Officer of the Bahrain International Circuit. 'His Royal Highness was travelling on Concorde to the US, Sir Jackie Stewart had just started Stewart Racing, and they were both on the plane. He stood up, said hello to Sir Jackie and said, "Congratulations on the team, and good luck."'

Unbeknown to Stewart, the Crown Prince was a motorsport aficionado, and had grown up watching Stewart and his contemporaries in Formula 1.

'Sir Jackie didn't know the Prince, and then they sat down and they started talking and the Prince is like, *I want to build a racetrack.*'

Stewart issued an invitation to attend Italy's grand prix, at Monza, in 1999, and there the Crown Prince was introduced to several leading figures. Most importantly he was introduced to then Formula 1 supremo Bernie Ecclestone, who controlled anything and everything to do with the championship – most importantly the finances – from the early 1980s until 2017. After cursory meetings, architect Hermann Tilke – who was increasingly the go-to man for new venues, having recently constructed Malaysia's new Sepang Circuit – joined the project and Bahrain officials set about planning the facility. An area of land in Sakhir, halfway down the western side of the island, was selected on which to build the circuit.

'This was going to be done in three phases: the inner track, the outer track and then the Formula 1 track,' explains Sheikh Salman. 'And when the Crown Prince went to his father, the King, he was like, *This is what we're doing*, and the King is like, *Yeah, but you want to get [Formula 1] one day, just build it.*'

The King approved plans to construct the facility in one go, at a cost of $150 million, in spite of Bahrain not yet having a deal with Formula 1.

Nonetheless, advanced talks were underway behind the scenes and a contract was eventually signed in September 2002.

Formula 1 was in an era of seeking to expand beyond its European heartlands and Bahrain's interest came at a convenient time. The commitment displayed by the Bahrainis convinced Ecclestone that this was a serious – and importantly for Formula 1, financially beneficial – long-term project, not a flash-in-the-pan vanity project.

'When Bernie said: "OK, why am I coming to Bahrain?", the story is, there was a map,' Sheikh Salman relays. 'And he was told: "You are the world championship right now. You're flying to Japan, Australia, and you're flying over this part of the world where up and coming this will be a player. And we should have a race on the map here." And Bernie said, "That's the middle of the water." "Yeah, there is an island there on the map!"'

Formerly reliant on its pearl industry, oil-rich Bahrain wanted to raise its profile on the international stage.

'If you're going to put a country on the map, sporting events are used,' says Sheikh Salman. 'But then if you choose the Olympics or the World Cup, that's every four years. And once you have it, when is the chance of you having it again? Whereas Formula 1 is every year.'

In an arid and rocky landscape, the enormous construction project proceeded, while the country also got ready to welcome Formula 1 by modernising its infrastructure, building new hotels and highways to accommodate personnel. Sandstorms and rare flash floods interrupted the build, though there was still time on Bahrain's side, given that the first event was planned for October 2004. Then came a development in late 2003 – Bahrain's first race was brought forward to April, giving no leeway for setbacks.

From groundbreaking to the completion of an international sporting facility took just 486 days, and Bahrain got the circuit finished in time, a couple of weeks before the inaugural event.

The 'keys' to the Bahrain International Circuit were handed over to King Hamad Bin Isa Al Khalifa on 17 March and Formula 1 cars took to the track on 2 April. This was the first grand prix to take place in a desert, and to combat the threat of sand the track surface and its surroundings were layered with a non-stick substance. At the same time, 1100 palm trees were installed around the facility to add to the ambiance and these are wrapped in tiny lights to provide illumination at night.

This was a brand-new culture, and region, for Formula 1, which sought to explore new terrains, most notably ones with financial incentives: the near-bottomless pit of oil money available in the region promised to boost Formula 1's coffers.

Bahrain has firmly established itself on the schedule. It first hosted the opening round in 2006, when Melbourne was preoccupied with the Commonwealth Games, and returned in 2010 when an elongated but unpopular circuit layout was used to mark Formula 1's 60th anniversary. Bahrain hosted the opening round from 2021 to 2024, and is due to do so for the foreseeable future through the rest of the 2020s and 2030s, outside those years when the timing of Ramadan will make this inconvenient and the honour will temporarily revert to Melbourne. It means that those usual first round tensions, nerves and expectations are increasingly fostered in Bahrain, rather than Australia, and the country's footprint is such that it is no longer the new kid on the block.

'Now, with 20 years under our belt, you talk about the history of F1 and we like to say that we're a modern classic race because there's a lot of history in F1, which we respect,' says Sheikh Salman. 'And we're never going to compete against the Monzas and the Spas, and yet with 20 years now, I think we have a history as well.'

The facility still feels fresh and modern despite it now beginning its third decade. The 10-storey Sakhir Tower dominates the skyline of the facility, located on the inside of the first complex of corners, just beyond the turnstiles of the paddock gates. The expansive paddock is welcoming, with a throng of palm trees cutting through the middle of the concrete. Illuminated at night, they brighten up the sand-coloured hospitality suites and pit building on either side.

Bahrain is not the most challenging circuit on the schedule, but the abrasive circuit is tough on tyres – especially rear tyres – meaning careful management is required through the race so that they aren't overly stressed or overheated. The circuit also has several medium-to high-speed corners where finding a comfortable balance is essential: the venue is susceptible to gusty conditions as winds blow in across the sea and the empty terrain. Drivers regularly use the national flag atop the Sakhir Tower as an indication of the wind's strength and direction, aware of how a headwind, tailwind or crosswind will impact the car's behaviour. The trickiest corner is Turn 10, a downhill left-hander that tightens on apex to a near-hairpin and where drivers frequently lock the front-left tyre, under braking while rotating the car.

Seven-time World Champion Michael Schumacher became the first victor of Bahrain's grand prix in 2004 – Turn 1 at the circuit is now named after him – while the likes of Fernando Alonso, Jenson Button and Sebastian Vettel also triumphed through its first decade.

Bahrain's race truly kicked into another gear when it became a twilight race in 2014, shifting the event into the cooler evening conditions. The cars came alive beneath the floodlights, facilitated by a frantic scrap for the lead between Lewis Hamilton and Nico Rosberg. The Mercedes pair swapped places on several occasions, battling wheel to wheel just inches apart, and their rapid pace – at times seconds faster than the opposition – firmly underlined the scale of the car's performance. It was the first race at which the Hamilton/Rosberg rivalry truly began to bubble, as Hamilton defeated Rosberg to become the first winner of a floodlit Bahrain race.

In 2020 – with the event rescheduled for November due to the pandemic – it was the scene of Romain Grosjean's violent accident and astonishing escape. Grosjean was involved in a first-lap accident in which his Haas VF-20 speared through a crash barrier and was sliced in two. Upon impact the car erupted into a fireball and the survival cell, containing Grosjean, was stuck within the mangled barrier. Remarkably Grosjean survived almost half a minute in the burning wreckage, then was able to extricate himself from the car and leap to safety.

'I mean, every time you see the footage, the pictures, it's amazing,' said Grosjean. 'If you take anything in that instance, like, slightly differently, I may be dead. There was fire all around, but [it was like] there was no fire in the cockpit. Obviously, there was a lot of fire in the cockpit! But yeah, I didn't realise that. When I jumped, I remember jumping the barrier, thinking I went off the track and I ended up being on track. And I was like, *Why am I on track?!* Obviously it's a night race and there was a lot going on, so I had no idea where I was, but it didn't really matter where I was. What matters is, like, *How do I jump out?!*'

Bahrain has been absent from the calendar only once since its inaugural event. The country was affected by civil unrest as part of the Arab Spring of 2011, when a wave of protests swept across the region. The grand prix was called off a few weeks in advance, marking the first time in several decades that a race did not take place. Bahrain's grand prix returned in 2012, amid heightened security, and there remains criticism over Formula 1's presence in the country.

'I think there's an opportunity with any global sport, whether it's F1 or whatever, and I think it's right to highlight the bad stuff,' says Sheikh Salman. 'We don't like to shy away from that or hide that. I think it's the right thing. And like any platform, this is where it should be discussed. I remember in 2011, it was a big thing for us and I think that was a blessing in disguise. We had to own up and say, *Yes, there are issues and yeah, we'll solve them and we'll fix them and we'll continue.* If that never happened, no one would highlight these things even in the region. And I think Bahrain is a test, a pilot study for the region.'

Bahrain – whether it is the first race or not – is committed to Formula 1 on paper until at least 2036 and in reality Bahrain will continue in the championship long beyond that date. Formula 1 is a prized asset for Bahrain, which still has a small footprint in global sports, and the country has deep pockets to ensure finances are no issue.

'I remember people always talk about this like you have the longest contract or you have . . . whatever,' says Sheikh Salman. 'We are in this for a relationship and the long game. It's not a five- or ten-year deal. It's not something we like or we don't like. We are committed on all levels.'

That commitment has also been seen elsewhere in Formula 1. Bahrain's sovereign wealth fund Mumtalakat bought into McLaren Group, the parent company of McLaren's Formula 1 team, in 2007, and became the sole owner in 2023.

And getting a head start – with Bahrain's grand prix now into its third decade – undoubtedly facilitated the country's international growth.

'The long contract definitely helps us,' says Sheikh Salman. 'We're a small island in a region where our neighbours are a lot richer than us. We started early, and I always laugh; can you imagine us trying to get a Grand Prix today after Saudi, after Qatar, after the fees? No way. They wouldn't even [look at us]. If we weren't the first in the region, there's no way we'd be a player. That's vision. You can't put a weight on that. You can't put a value on that. Any good idea, when you start and you say, *Yeah, how did that work out?* You know, imagine trying to do it now, forget about it.'

There is also healthy competition within the region as motor sport continues to grow.

'A lot of people said when Abu Dhabi came [in 2009], *Are you worried?* I'm like, *I love this. I love competition*,' says Sheikh Salman. 'And we have a head start on building skill sets, human capital, all the experience and I think with Saudi coming on board it was such an important thing because of the fan base in Saudi, plus their population is 70 per cent below the age of 30. That's huge for us. And the Eastern province [of Saudi Arabia] is 20 minutes away.'

Formula 1 is comfortably the biggest sporting event for the tiny island, which is immediately obvious at the recently renovated airport: decorated with Formula 1 race posters, it has special lanes upon arrival for personnel. The teams are also highly visible, travelling together and dressed in specific travel gear with branded luggage. The Ferrari personnel are always the most immaculately presented, on account of their crisp black blazers, complete with Prancing Horse emblem. (Italians always find a way to bring the style.)

That sense of anticipation and enthusiasm is heightened when Bahrain is the opening round, with season passes requiring collection, everyone regathering their bearings in the paddock and catching up after the winter break. Unfortunately, the atmosphere is still a little sterile even after two decades: despite a blend of Western expats and curious locals, Bahrain's record attendance stands at about a fifth of the best-attended grands prix. There are swathes of the track where there are no grandstands or spectators, making it feel more like a test day than a round of the world championship. And yet, in a population of 1.6 million, attendance figures of 100,000 for the weekend and 37,000 on race day are encouraging, and interest inside the country is gradually growing. Once Bahrain's event is complete it's straight on a plane to hop across the region to one of Formula 1's newest rounds.

Formula 1's first venture to the region, in 2004, was to Bahrain

6

SAUDI ARABIA GRAND PRIX: JEDDAH CORNICHE CIRCUIT, JEDDAH

> *'You need to be super precise and that makes it very, very difficult, because if you are out by 5 or 10 centimetres, it's not like you go wide and that's it. You touch the wall – and it's done!'*
> CHARLES LECLERC

Considering Formula 1's ruthless pursuit of cash, the constant striving to increase revenue and Saudi Arabia's plentiful reserves, it seems remarkable that the parties took until the early 2020s to unite.

Saudi Arabia has aggressively opened its doors to the world in recent years and Formula 1 has slotted into that strategic framework.

In 2016 de facto leader Mohammed Bin Salman started Vision 2030, designed to diversify Saudi Arabia's economy, soften its international image, establish a tourism industry and reduce its long-term dependency on oil.

Sport is a major component of Vision 2030, impacting major international competitions in a breathtakingly short space of time, as an insular nation opened up, spent with abandon and hoovered up global championships. Saudi Arabia has limitless pockets and primarily achieves its ambitions through its Public

Investment Fund (PIF), believed to be among the richest sovereign wealth funds.

PIF has exerted its influence in football – through club ownership and league expansion – in golf, in tennis, in boxing – among others! – and is striving to host major blue riband events, such as the World Cup, which is due to be held there in 2034, and is expected to bid for the Olympics.

The state-owned oil company Aramco, which reported annual profits of $121 billion in 2023, became a global partner of Formula 1 in 2020, its signage prominent trackside at grands prix. In 2022 Aramco became the title partner of Aston Martin, meaning the team is officially known as Aston Martin Aramco, a deal that was subsequently extended through 2028, and it has forged a close alliance with the team over several technical projects.

Inevitably Saudi Arabia wanted a grand prix of its own and in November 2020 a 10-year deal was announced, starting in 2021, for an event that takes place at night. A wall-lined track, designed by Carsten Tilke, was laid out along the corniche of Jeddah, Saudi Arabia's second city, which is located halfway along its vast Red Sea shoreline.

Such was the rapid turnaround between the announcement of the deal and the first race – which was installed as the penultimate round of 2021 – that there were doubts the facility would be ready in time. Most of the existing roads were remodelled, giving designers additional freedom when it came to its layout, while swathes of the 27-turn circuit, along with the required infrastructure, had to be constructed from scratch. Some who visited the facility a few weeks out from the inaugural event for a recce were gobsmacked at the volume of work that was still necessary. The paddock resembled a building site and there were still a few incomplete areas on the site when Formula 1 arrived: a few paths that led nowhere, a couple of buildings still under construction and the advice not to lean on surfaces lest they had been recently painted. Nonetheless, everything essential for the smooth operation of the event was completed in time and in December 2021 – just over a year after the announcement of the grand prix – the Jeddah Corniche Circuit played host to a frenzied race. Jeddah's debut affair featured two red flags and a litany of accidents and incidents, including at the very front, as the title battle between Max Verstappen and Lewis Hamilton reached a furious peak. Verstappen and Hamilton hurtled around Jeddah as if competing in a different category to their rivals, and traded paintwork as they scrapped, before Hamilton eventually prevailed, putting the title contenders level on points after 21 races and in the process becoming the first victor of the Saudi Arabian Grand Prix. This marked the only time that Jeddah played such an influential role in a championship showdown, as from 2022 it was moved up the calendar to the second event of the season, establishing itself in March. Verstappen won that race, defeating Charles Leclerc, while in 2023 Sergio Pérez made it three different winners from the first three races, before Verstappen became the first multi-time Jeddah winner in 2024.

It is a high-speed venue, with walls perilously close by, and drivers need to quickly find the flow, and have confidence in the car, in order to perform well. From Turn 4 through to the hairpin at Turn 27, drivers will rarely flick below fifth gear, with a sequence of rapid changes of direction, blind corners that have raised questions over safety and no room for error. Get it wrong and it can go very wrong. In 2022 Mick Schumacher lost control through Turn 10 and crashed heavily during qualifying, and the damage – and concerns over spare parts moving forward – was such that his Haas team withdrew the car from the following day's race. The average lap speed of 254km/h (158mph) during qualifying makes it among the fastest circuits on the calendar.

'You need to be super precise and that makes it very, very difficult, because if you are out by 5 or 10 centimetres [2–4 inches], it's not like you go wide and that's it. You touch the wall – and it's done!' says Charles Leclerc. 'So to find the confidence on a track like this is much more difficult than other tracks.'

It means drivers have little room to relax through the course of a lap.

'The grip is pretty high for a street circuit,' says Daniel Ricciardo. 'The first sector is super fast: once you get through Turns 1 and 2, it's time to hold on because the G-forces on your body are very high. It's a real track, very physical.'

The Jeddah Corniche Circuit itself is an elongated venue, adjacent to the Red Sea, around 10km (6 miles) north of the very centre of the city. The knife-shaped track cuts its way northwards for almost 3km (2 miles) before turning back on itself at Turn 13, a banked turn of 180 degrees, behind which is the marbled Al Rahmah mosque that sits on stilts over the water, and the modern Jeddah Yacht Club and Marina, giving a slice of Mediterranean feel in the Middle East. The final sector, with a long, sweeping curve that eventually tightens into a hairpin, encircles a man-made lagoon, on which the Media Island was constructed.

There is a peaceful, parklike feel to the island, shattered occasionally by the hum of a Formula 1 engine, the Saudi Falcons aerobatic display team practising in their Hawks over the Red Sea, or the distant construction noise in an ever-evolving area. When the sun sets the island is illuminated by a plethora of green and blue lights that shimmer across the lagoon while bright white spotlights beam towards the sky. A spectacular drone show takes place before and after the race, manoeuvring into positions to create team logos, racing cars and countdown clocks.

It is not a grand prix with the highest attendance – around 140,000 enter the facility across the course of the weekend – and there is a large focus on premium hospitality. A spectacular U-shaped building sits over the circuit, providing views of Turns 1, 2 and 22 – where most of the action takes place at the end of the pit straight – while there is also a viewing platform towards the Red Sea, a firm example of what money can build. The expansive nature of the Jeddah Corniche Circuit means golf buggies zip around the venue while at prime times there is a constant flow of vans with blacked-out windows ferrying the VIPs and the VVIPs.

Expanding that fan base is clearly a target of Saudi Arabia's officials across the coming years.

'What do we need? How do we overcome our challenges? How do we move forward towards a better future for everyone? And this is across the board in every sector,' said Minister of Sport Prince Abdulaziz bin Turki Al Saud during a media roundtable in 2023. 'Sport is one of the biggest pillars within Vision 2030. If you look at our numbers, in 2016, 13 per cent of Saudis participated for half an hour or more in sports. In 2022, we reached 48 per cent. Our target for 2030 was 40 per cent. So we exceeded that target and now we're readjusting the KPIs [key performance indicators] that are on us to reach to.'

The headlines understandably focus on Saudi Arabia's trophied acquisition of major sports and events, but further down the chain there is a greater focus on grassroots competition.

'Sport is not just about participation, it's about a lifestyle, and it's more of a quality of life that you want to enhance: it's a big part of the 2030 Vision, and of course these events help us achieve that, and that's why we're hosting these events,' says Prince Abdulaziz.

Jeddah is a city that highlights Saudi Arabia's acceleration towards the future while counterbalancing with its past – a dynamic witnessed on landing at the gleaming international airport, where its arrivals hall is bedecked in adverts and branding for that weekend's grand prix. Mixing with the westerners, the seasoned travellers dressed in branded gear, are men wearing the white robes of Ihram that indicate they are embarking on umrah, the secondary pilgrimage to Mecca, for which Jeddah is the principal gateway. Each respect the other with a mixture of intrigue and curiosity, travelling for wholly different purposes. There is a high-speed train connecting Jeddah with the holy cities of Mecca and Medina, but within Jeddah there is a lack of public transport. There are plans to develop a metro system, but the city's high groundwater level has led to added complications and subsequent delays. Consequently driving is the way to travel around Jeddah – with petrol extremely cheap, at about 50p a litre – meaning that roads are regularly clogged. Tailgating, bewildering lane changes and speeding are commonplace on the stretches of roads that aren't congested. Road closures and diversions, especially due to the presence of 'The Formula', as locals call it, and the ring of security implemented for the event, can also complicate matters, such that missing a one-way road can easily add swathes of time to a journey. It is quite a contrast watching Formula 1 drivers thread the eye of a needle before heading outside to embark on the free-for-all in rush hour in Jeddah. There is nevertheless a slight feeling of detachment: the Jeddah Corniche Circuit is a trek away from the centre of the city and its UNESCO World Heritage Site, while the abundance of international hotel chains near the circuit means it's feasible to visit the grand prix and never go anywhere near Jeddah itself.

Formula 1 now slots into an ever-expanding motor-sport portfolio in Saudi Arabia. Formula E (2018) and the Dakar Rally (2020) set up shop in the country, while Extreme E – now Extreme H – has held a round in the desert since 2021. PIF already has deep ties with Formula E and Extreme H as part of the Electric 360 portfolio, the World Rally Championship will debut in the country in 2025, while MotoGP is also expected to land in Saudi Arabia this decade.

Long term, there are plans to move Saudi Arabia's grand prix to Qiddiya – a giant sports and entertainment hub under construction outside the capital Riyadh, which will be like Disneyland on steroids – but if that happens it will be in 2027, at the earliest. Under the proposals the Qiddiya Speed Park Track will nestle within a theme park, blending the best elements from a street track and more open course, and featuring an elevation change of over 100m (330ft) across the course of a lap. There is also due to be an elevated section, a 20-storey rise dubbed The Blade, while the Tuwaiq Mountain range will provide the backdrop. Qiddiya was initially anticipated to be ready as early as 2023 or 2024 and consequently Jeddah was intended as a short-term venue for Formula 1. As a result, ahead of 2024's grand prix, the entire paddock internals were refitted and upgraded, while Jeddah has increasingly established itself as the home of Saudi Motorsport.

'Jeddah will remain in place as a circuit, it's permanent – we call it a street circuit but it is an operating track 365 days a year, so it will remain a racing circuit with all the recognised infrastructure around it,' says Saudi Motorsport CEO Martin Whitaker. 'We have product launches, corporate events and we already do a lot of community events – it's open daily for running and cycling.

'It's nice to see where're we've gone from, in November 2021, to where we are today, where it's a real going concern. Any F1 race anywhere in the world is a platform on which you can create awareness, profile, do a whole load of activities. There's no doubt about it, in terms of the average fan and spectator, the interest has gone up. We have a karting centre and we're seeing so many young kids coming now, they want to go, they're interested in racing.'

That is a process which will not happen overnight.

'I think we are raising more and more awareness around here in Saudi with the younger generations to get into driving and racing,' said Max Verstappen, after winning in 2024. 'And I think a lot comes down to education and guidance. The more help and support that there will be from people with a lot of experience, the more that you can grow it here and the passion. So that takes time. It's not that it happens within five years and suddenly there will be a Saudi driver suddenly coming through the ranks. You need to really work on that. It's good to see that they're loving motor sport and you can see the fans here. They're quite young in general, what you see around here.'

Jeddah's grand prix is not universally loved, but Saudi Arabia's involvement in global sport is only going to grow at a searing speed as it flexes its financial muscles. Formula 1, chasing the money, is going to be a key pillar of that expansion.

Each Formula 1 team is a well-drilled machine when it comes to the race-day schedule. Every member of the roughly 60-strong team undertakes their specific duties to add their piece of the puzzle to the jigsaw, fettling and tweaking certain components across the car. There will be a Team Manager, overseeing the entire operation; a Chief Mechanic ensuring everything is shipshape inside the garage, managing the components and liaising with different crews; and a Chief Race Engineer, whose job it is to form a strong bond with a driver, and be their eyes and ears during a race.

If anyone is slightly out of kilter, or misses something from a checklist, then the entire operation could be put at risk. But everyone trusts each other to get on with the job. The car must be ready to leave the pit lane 40 minutes before the start of the race.

At each circuit there is a pits complex, the usual design being an enormous building that contains a row of garages on the ground floor to house the cars. These are divided between the 10 teams, and there are also a couple for the safety car and medical car. Outside of each pit garage is a small space for various paraphernalia and spare parts – most notably front wings – and of course the pit box, each team marking out their area for drivers to undertake their pit stops. They filter in from and into the fast lane of the pit lane, and between that and the circuit is the pit wall, which features temporary structures. The pit wall has an array of screens, communication methods for senior management to watch during on-track sessions, and seats – colloquially called 'prat perches'.

Saturday into Sunday is the only night of the weekend without a prescribed overnight curfew for the mechanics – earlier in the event, the hours at which personnel are permitted to be in the paddock are restricted to protect their wellbeing – but now the cars are tucked up under the covers in their respective garages. It means Sunday morning can be among the quietest time for mechanics, compared to Wednesday through Saturday, when cars are constantly being built, maintained and fettled.

Once the qualifying session begins, usually at 16:00 local, the Formula 1 cars enter *parc fermé*, meaning they are now off limits to team members, except under strict supervision. Their set-ups and specifications cannot be altered again for the remainder of the weekend, and they must qualify and race with the same specification of component, and the same set-up. The same type of component can be changed between qualifying and the race, but if there is a change in specification – a driver damages the only specification of that front wing in qualifying, for example, and must revert to a different version – they must start the race from the pit lane, meaning they will join the race at the back of the pack.

The translucent covers must be applied to the cars two hours after the conclusion of qualifying, with a seal applied by the FIA, and they can be removed only five hours before the start of the race.

On race day morning the mechanics will make checks to the car, ensuring all the components are in order as well as undertaking a stocktake of the spare parts and checking that the various tools are in the right drawers. The car will eventually be fired up to enable reliability checks as well as to put the fuel in the car, and various approvals from the FIA must be secured. Mid-race refuelling has been banned since 2010, meaning the cars start the races with around 110kg (242lb) of fuel and the fuel flow rate must be managed throughout the course of the grand prix. This is another area where teams can strive to seek an edge and fuel suppliers strive to produce the best fuel within the prescriptions of the FIA regulations. From 2026, as part of Formula 1's sustainability push, the championship will rely on fuels that are 100 per cent sustainable. Around half of the grands prix teams will also be cognisant of the support races ongoing through the morning: Formula 2 and Formula 3 use the same pit boxes in front of the garages and temporarily commandeer the pit walls too.

Drivers will arrive a few hours before the race and one of the most important pre-race elements is the strategy meeting. Each team has supercomputers that simulate thousands of potential race strategies depending on the circumstances: the extent of expected tyre degradation; the benefits or pitfalls of the undercut or the overcut; the position of the drivers after the first lap, factoring in their grid positions and the risk of an opening corner collision; the simplicity of overtaking versus the importance of track position; the risk of a safety car or virtual safety car interrupting proceedings; the threat of weather; the pit lane loss time; and so on. Each driver will have a primary strategy and various alternative strategies depending on how the race unfolds, and these will often be coded to ensure that rival teams remain as in the dark as possible. At some events the strategy may be fairly straightforward, but at others there may be more variables to consider. A team towards the front of the grid will want to adopt the simplest, most straightforward strategy to collect maximum points, while those towards the rear of the field may try unconventional strategies, or even gamble on unexpected happenings, in order to hope that the race will fall their way.

Drivers must be aware of their main strategy, but also their potential back-up strategies, while strategists must also be reactive to anything that occurs, in the event that a driver needs to be called in for a pit stop just seconds before they pass pit entry. Clear and proactive communication between the drivers and the various personnel in senior positions – predominantly a driver's race engineer – is essential.

These relationships build up over time, and a weekend's worth of track activity and meetings means that everyone – in theory – should be on the same page come Sunday. Drivers will regularly walk each track on Thursday with the engineers, not just to get the lie of the land but to note any potential changes, while also getting an hour in a more relaxed atmosphere away from the paddock.

Drivers may also have meet-and-greets, interviews on the fan zone stage, or marketing and sponsorship duties – which usually take place in the exclusive paddock club suites above the pit lane, their extent varying depending on the location of the race. They take lunch in and around the drivers' parade, which takes place two hours before the race. The crew will have lunch in the motorhome or, more usually, in the pit garage, and get suited up in the fireproof overalls that they wear throughout the race. There may also be a rousing speech from a senior team member, and fist bumps between drivers and crews, though everyone by now knows the plan for the race ahead and the expectation of executing it to perfection.

As the mechanics prepare for the race and draw up their fold-up chairs in the garage – in the space created now that the cars are out on track – attention turns to the pit stops.

Pit stops are a crucial part of any grand prix and these are practised in the factories on mule cars away from events, and on each of the four mornings at grands prix – including on race day.

In a sport where every 10th of a second is being chased in terms of car performance, the pit stop is a place where races can be won or lost. Pit boxes used to be perpendicular to the pit lane but in recent years have been ever so slightly angled, after one team realised that this was an easier entry for drivers – and therefore faster – and rivals quickly copied the system. The team's chief strategist determines the laps for the pit stops and accordingly radios to the crew that it is time for them to take up their position in the pit box. They must always be ready to pounce, in the event of an unscheduled stop – perhaps because of a puncture – or if a sudden safety car deployment makes a pit stop viable.

There are usually 18 mechanics involved in the pit stop process and in recent years teams have begun to undertake proper warm-up sessions for the involved personnel, alongside the mandatory safety equipment. There is a front and rear jackman, to hoist the car off the ground, while a back-up stands both front and rear in the event that there is a failure with the primary component. The front jackman stands in front of the car, lifts it off the ground and then immediately steps to the side, from where he can lower the car and swing the jack out of the way when the stop is complete. At the back, the rear jackman stands to the side and pounces once the car is clear, to insert the device underneath the diffuser so that the stop can begin.

There are three mechanics on each of the four corners of the car: one to remove the old tyre; one to attach the new tyre; and one to detach and affix the wheel nut with the extremely noisy pneumatic wheel gun. The tyre gunner tracks the motion of the car as it enters the pit box and the process of detaching the wheel begins before the car is completely at rest. Once the wheel nut has been removed, the gun automatically switches such that the next motion tightens the nut back up for the new tyre.

Tyre off and *tyre on* may seem simple, but the tyres are heavy, at 9.5kg (21lb) for the fronts and 11.5kg (25¼lb) for the rears, meaning the mechanics must be extremely fit and strong. They must also perfect the slightly awkward motion of extricating the tyre from the rim at an angle and then gripping hold of the tyre after it has been removed from the car, to ensure it doesn't bounce away into the fast lane of the pit lane. As *tyre on* has started their motion before *tyre off* has been fully complete, any hindrance during this process will shave a knock-on effect. The tyres are kept in their protective blankets – to hold them at the ideal temperature – until shortly before the pit stop.

There are also two mechanics on either side, who hold the middle of the car to help stabilise the machine during the pit stop process. They can also quickly undertake duties such as cleaning the radiators or wiping a driver's visor if needed. There may also be mechanics poised at the front of the car to make a front flap adjustment, leaning across the front wing, if a driver wants a minor tweak to either facilitate or combat oversteer or understeer, made possible via a screw on the front wing.

In days gone by there was a lollipop man to signal for the driver to go, but this has been superseded by an electronic lights system above the driver's head, which turns green when the wheels have been fitted – though there is also someone on standby to the side of the front of the car in case of a busy pit lane, keeping an eye on the traffic, in the event that the driver needs to be held in the box. From the vantage point inside the cockpit, a driver cannot see a rival who may be in the fast lane or about to enter the box ahead.

Drivers must also be perfect when they stop. The position of the mechanics is drawn out on the pit box to the millimetre, and any incorrect car positioning will slow down the whole process, raising the prospect of a mistake due to personnel having to reshuffle their respective places. Drivers have been known to overshoot the pit box, sending their front jackman flying backwards, which is no easy ride considering the cars are still travelling at a rapid 96km/h (60mph) in the pit lane.

Nowadays any stop over 2.5 seconds is considered slow and anything over three a relative disaster. The quickest ever pit stop was recorded by McLaren, at the 2023 Qatar Grand Prix, where all four tyres were changed in just 1.8 seconds. Finding those final tenths is the hardest part. And this is about not just the human element, but also the technological aspect, with teams striving to develop wheel guns that can undertake the process faster.

But there remains the trade-off between speed and consistency: the greater the pursuit of overall pace, the higher the chance of something going wrong. And it can go very wrong. Loose wheels and cross-threaded wheel nuts are disastrous, both for safety and race prospects. The latter impacted Valtteri Bottas during the 2021 Monaco Grand Prix and, no matter how hard Mercedes tried, the wheel could not be removed. The car had to be returned to the factory, two days later, before Mercedes could finally remove it. Haas suffered a double failure at the 2018 Australian Grand Prix, with cross-threaded wheel nuts in successive pit stops costing it the chance of having two cars in the top five positions.

Teams must also ensure they have the right tyres in the right place; in the rain-hit race of 2016, narrow garages in Monaco meant that Red Bull did not have the compound of tyres it needed in an optimum position, and the time lost in getting them to the pit box cost Daniel Ricciardo the victory.

On occasion, usually during a safety car period, teams undertake a double stack, meaning both drivers come into the pits on the same lap. That means teams must have the new tyres for driver 1 lined up first, followed by the new tyres for driver 2, with the high risk that a mistake for driver 1 will also cripple the prospects for driver 2 – or that the tyre sets will be mixed up. Each driver has their own allocation set by Pirelli, usually defined by teams with their initials inscribed on the sidewall, but at the 2020 Sakhir Grand Prix everything went wrong for Mercedes. A radio failure meant the mechanics were late getting into position and chaos unfolded. George Russell came in and took on a fresh set of tyres, then Valtteri Bottas entered the box to have fresh tyres fitted and Mercedes realised that it had only the 'GR' tyres. Russell was already out on 'VB' tyres, meaning he had to come in again, and amid the confusion Bottas was refitted with his old tyres. The net result was that Russell had to come in again to be fitted with the correct tyres, while Bottas had endured a near half-minute stop and was still on used tyres. A sure-fire 1-2 result had been lost.

The crew will typically have to undertake four pit stops during the race – if the race is a conventional two-stop strategy and both cars remain in proceedings – though they must remain alert even after the final scheduled stop just in case there's a late issue.

After the chequered flag, some mechanics will have the pleasure of running to stand beneath the podium, to celebrate with their driver – or drivers – while others will have time to begin pack down or to wait for the moment where they can collect their cars.

Engineers and senior personnel will now have a quick debrief, though more in-depth meetings are likely to follow – perhaps over a video call – in the downtime between events. This offers the opportunity for a greater understanding of what has unfolded, which is helpful after the immediate adrenaline and emotion has faded. Even the winners will strive to unearth aspects of the race weekend that could be undertaken in a better manner, to search for that additional hundredth of a second, or the small margins on which a race may swing.

The race now over, the cars are left in the world's most expensive car park in the pit lane – gradually cooling down, spattered with the detritus and scratches of the on-track combat – while the FIA undertakes their post-race checks. It is not possible for every component to be inspected, but sufficient parts are analysed to dissuade anyone from breaching the regulations. If any fault is found, the team in question will be referred to the stewards for further consultation. The worst-case scenario is something which is clear-cut, which means a disqualification is a slam dunk, wrecking the entire weekend's work. Such a situation happened to Mercedes' Lewis Hamilton and Ferrari's Charles Leclerc after the 2023 United States Grand Prix, when the underfloor plank had worn down too much – due to the bumps on the track and the cars being run too low – and no longer complied with the FIA's thickness measurements.

The same can be said if a driver is summoned for an incident during the race. Race Control will refer incidents to the four-person panel of stewards, and while some are investigated and determined during the race, consultations on some incidents will happen post-race. Drivers are usually accompanied by a team representative, typically the Team Manager, who will have an encyclopaedic knowledge of the rulebook and past incidents, in a bid to present the case for the defence. The stewards will eventually deliver their verdict, which could range from no further action taken, to a post-race time penalty or a drop of grid positions for the following round's race. Sometimes such matters, or more broad issues surrounding the sport, are discussed at the next event's drivers' briefing. All 20 drivers – along with reserve drivers and senior race officials from the FIA – attend the meeting each Friday evening, and they have been known to drag on into a second or even third hour if there is a lengthy agenda.

Eventually each car is approved to be released by the FIA and the teams put them on a tea tray and wheel them back to their pit garages, where they will be stripped down, their parts checked, and packed away for the transit to the next event.

And by the time Saudi Arabia's event is completed, following on from Bahrain, it's been nearly two weeks in the region, with everyone ultra-keen to scuttle to the airport to fly home.

7

MIAMI GRAND PRIX: MIAMI INTERNATIONAL AUTODROME, MIAMI GARDENS

'The role of the logistics departments – it's our challenge to create the exact same environment and working conditions in 24 different places across a year, regardless of temperature, weather conditions, the local environment, the local political state.'
PETE CROLLA, TRACKSIDE OPERATIONS
MANAGER FOR HAAS F1

Formula 1 now makes three trips to the United States of America – to Miami, Austin and Las Vegas – and the first annual voyage takes the championship to the southern tip of the Sunshine State and to one of the newest additions to the schedule.

Expansion in the United States has long been an ambition for Formula 1 and in 2010 there was the tantalising prospect of a New York Grand Prix. A circuit was mapped out around the streets of Weehawken, New Jersey, with the skyline of New York providing the backdrop. The event – provisionally known as the Grand Prix of Americas – was installed on the 2013 calendar, but it was later postponed until 2014 because the organisers were not able to get everything ready in time. It transpired that the promoters had been unable to raise the required funding to even get the project off the starting blocks and that proved an insurmountable hurdle, in an era where short-term gain was prioritised over long-term vision. The proposal slipped off the radar.

When Liberty Media acquired Formula 1 in 2017, Miami was swiftly pinpointed as a prospective 'destination' grand prix. This is a term coined by Sean Bratches, Formula 1's Commercial Director from 2017 to 2020, to highlight how the sport can interact with a city's promotional needs. In effect, Formula 1 benefits by being so close to a major city – taking the championship to new fans and environments – while the location itself prospers from the influx of interest that Formula 1 brings and, of course, the collaboration is mutually beneficial from a financial perspective.

Proposals were drawn up to hold a grand prix downtown, in and around Biscayne Bay, but those plans stalled amid local and political opposition, notably concerns over the disruption caused in terms of the road closures and infrastructure required. That proved to be the catalyst for a fresh concept and by mid-2019 a new project was unveiled, which shifted the location to the multi-sport Hard Rock Stadium, the site owned by the group that was trying to bring Formula 1 to Miami.

The pandemic delayed matters further and some local residents lodged complaints, but these were addressed by promises to assist the local community, such as integrating businesses and restaurants into the campus, educational programmes for schools and the agreement that track sessions would not overlap with school hours. Once such hurdles were cleared, Formula 1 agreed a 10-year deal to hold a Miami Grand Prix from 2022.

Tentative plans to utilise some public roads in the area were shelved and organisers instead opted to keep the entire circuit within the confines of the Hard Rock Stadium's campus. Officials assessed and evaluated 45 potential iterations before eventually finalising a wall-lined street-style circuit, with the start/finish straight adjacent to the Hard Rock Stadium. A permanent three-storey paddock building was constructed in the shadow of the stadium and walkways between the two structures were completed in time for the second grand prix in 2023, allowing for the potential Formula 1's garages and facilities to be utilised at the campus through the year, such as tailgating in the pit lane during football games. It is a moderately interesting circuit, given the limitations of the flat terrain and the campus' facilities, but not one that will truly test the drivers or leave them enthralled.

'With the Miami campus we have had a lot of freedom and independence in ways that other circuits have challenges with,' explains Miami Grand Prix CEO Tom Garfinkel, who also holds the same role for the stadium and the Miami Dolphins. 'We are in a fortunate position where we own the real estate, it's not owned by the county or city. We wanted Miami to be a global destination by bringing the biggest events in the world here. With Hard Rock Stadium as a centrepiece, we had a clean sheet of paper to design a racetrack.'

Hard Rock Stadium is better known as the home ground of NFL's Miami Dolphins – which has facilitated crossover promotional material between Formula 1 and the NFL – and it has held six Super Bowls, most recently in 2020. Hard Rock Stadium also hosts tennis' Miami Open and will be a host stadium for the 2026 FIFA World Cup, and it's a concert venue for the likes of the Rolling Stones, U2 and Green Day. It dominates the skyline of the entirely flat surroundings, and the tall white spires in each corner and its signature white canopy – introduced as part of a remodelling in 2016 – make the stadium one of South Florida's most recognisable landmarks. Garfinkel joined the organisation in 2013, five years after businessman Stephen Ross bought a majority stake in both the stadium and the Dolphins, and the pair have evolved the campus.

'When I first came here, this was a 28-year-old stadium with a bunch of empty parking lots,' says Garfinkel. 'We've had Jay-Z, Beyoncé, Coldplay and U2. We've hosted the Miami Open tennis tournament, the Super Bowl, College Football Championships and the Rolling Loud Hip-Hop festival. I think we're the only place in the world that has hosted Roger Federer, Tom Brady, Messi, Neymar, Venus and Serena Williams, Floyd Mayweather and now Lewis Hamilton.'

The grand prix slots into early May as part of the Hard Rock Stadium's year-round usage as a sports and entertainment venue, in a city that has glamour and high-profile names in its DNA.

'We have a pretty tight window because of the other events that take place here at this facility,' says Garfinkel.

The NFL main season runs from September through early January, with the Miami Open held across two weeks between late March and early April. That means detailed planning is required to ensure everything runs without a hitch.

'We first start looking at the logistics of moving equipment in and out of Hard Rock Stadium about 10 months in advance,' says Todd Boyan, Senior Vice President of Stadium Operations. 'The biggest part of this operation is coordinating between all the different entities to make sure we understand what needs to be installed and when. The key is not to install something which then hinders another piece of the puzzle.'

The inside of the Hard Rock Stadium was effectively an underutilised storage facility at Formula 1's first event in 2022, when the Miami Open's centre court remained fully erected inside. But from 2023 the paddock was relocated on to the football field, creating a communal square dubbed the Team Village. This meant new plans were needed. Now, the stadium's turf is pulled up after the Dolphins' last home game and a protective flooring is applied. (And yes, there are back-up plans in case additional fixtures are required in the event of a play-off run.)

Barrier and grandstand construction begins shortly after the conclusion of the NFL regular season, with around 40 per cent completed by the time the Miami Open begins.

'We have to be strategic in the way we carry on the construction and lead-in for Formula 1 so that we ensure guests coming to the Miami Open continue to have a terrific experience at a world class tennis tournament,' Boyan stresses.

Work continues during the Miami Open where possible and the remainder – usually the final 40 per cent – takes place through the rest of April.

'Taking down the Miami Open tennis court is normally achieved within a 40-day time frame,' says Boyan. 'We removed it in 12 days [for the Team Village]. That allowed us to build the 13 hospitality structures a couple of days earlier than planned for the F1 teams, on what was once the Stadium Court and, before that, a football field less than four months ago.'

The creation of the Team Village allowed spectators to enter the 300 level of the Hard Rock Stadium – the top tier of the stadium – and look down on the paddock from above. The spectators are still a distance away from the drivers, but this marked an opening up of the inner sanctum, which is usually hidden from view, and makes it possible for spectators to hear – at least briefly – and cheer. It also provides a panoramic view of the campus – and therefore the circuit – from the back of the stadium.

'When you start getting to the point where drivers are walking across the paddock and people are cheering or chanting, that's authentic,' says Miami Grand Prix President Tyler Epp. 'Now we're building something, something authentic to Miami, we have something – that takes time and it grows.'

Formula 1 rolls out of town on Sunday night straight after the race, but for the team at the Hard Rock Stadium there is still a strict protocol to follow.

'When the teams leave, we'll start to deconstruct the Team Village,' Boyan explains. 'We won't put the turf back in just yet since the summer is when we carry out the bulk of our entertainment events, which in the past has included major concerts, boxing and international soccer. The grass pitch will then be installed in early August in preparation for our first Miami Dolphins game of the season. As for the temporary structures and grandstands for Formula 1, it roughly takes about six weeks to dismantle everything. We're learning to be more efficient each time we do this, so next time the process will be even quicker.'

Compared to the likes of Melbourne, Monaco or Singapore, the Miami moniker is a slightly looser affiliation. Miami – and Miami Beach – has the white beaches, the art deco buildings, the relaxed vibe and the influence of several cultures, given its location at the crossroads of the Americas.

'It's really close to home,' says Mexico's Sergio Pérez. 'It's a two-hour flight, so a lot of friends [are] also here every year. I used to come two or three times a year to Miami, so now having the race, it's really nice. I get to see some of my friends that are living here. It's a city that I enjoy a lot, being here.'

The Miami International Autodrome is actually in Miami Gardens, 24km (15 miles) north of downtown Miami – and a 90-minute journey when Interstate-95 is at its worst. Part of the circuit runs underneath the concrete access ramps of Florida's Turnpike and just across the canal, which is known as Snake Creek, are residential roads.

That would risk being a limitation, but Miami organisers have been savvy enough to connect the Miami International Autodrome to the city, to local artists, musicians and restaurant outlets.

'A lot of what we do at the circuit is representative of what the city and of South Florida has to offer,' says Epp. 'Miami is our brand, and we try and bring that out with the circuit.'

The campus' grounds were heavily leveraged to maximise premium hospitality areas around the venue, while leaning into existing facilities: its glass-floored gondola zips above the track along the back straight and permits spectators to look directly down at moving Formula 1 cars. One of Miami's constructs quickly captured widespread attention: the MIA Marina. Installed on the inside of Turns 7/8, it features yachts surrounded by sand-coloured carpets, on vinyl purporting to be water, giving it the nickname 'the fake marina' – all part of the campus-wide creativity that doesn't take itself too seriously.

Miami's first grand prix in 2022 was won by Max Verstappen – who doubled up with another win in 2023 from ninth on the grid – and the crossover was in evidence on the podium: the top three donned NFL helmets during their walk-on. Verstappen's spell on the race was broken in 2024 when McLaren's Lando Norris ended his drought in fine style, beating Verstappen in a straight fight to pick up his maiden victory.

The race has already earmarked itself as a celeb-fest, attended by the likes of Tom Brady, Michael Jordan and David Beckham, and was the place to be seen that year. That gave Miami instant impact, even if more hardcore fans bristled at the slide towards showbiz at a circuit that is far from the most exciting. The key for Miami in the long term is to retain that showbiz sparkle, particularly given that the newness factor has worn off and Las Vegas is now ostensibly the glitzier and higher profile race. The early signs have been encouraging, with interest maintained and spectator numbers inching upwards year on year, in part thanks to Miami organisers managing its campus capacity.

Miami's position in the schedule means it is now the final of the early season 'flyaway' events, which now stretches to six grands prix, across the Gulf, East Asia, Australia and North America. It is a busy schedule and means that Formula 1 logistics have to be pinpoint perfect.

Formula 1 travels to 24 circuits per year, across 20 countries on five continents, and while approximately 3000 people have to get around the world – travelling over 160,900km (100,000 miles) annually on roughly 50 flights – so too does all the equipment, from the cars to the computers to the starting lights, essential to the running of a grand prix.

'The role of the logistics departments – it's our challenge to create the exact same environment and working conditions in 24 different places across a year, regardless of temperature, weather conditions, the local environment, the local political state,' says Haas F1 Team Trackside Operations Manager Pete Crolla. 'We've got to be as consistent as we absolutely can. You're never going to be 100 per cent from one race to another. But if we're in the 90s, then we've done a bloody good job. It's really a group effort. It has to be a team effort because the scale of the operation, the amount of stuff that we move around and the speed at which we have to move it around at, is phenomenal.'

Formula 1 has been working with DHL for almost 40 years and the company became the championship's Official Logistics Partner in 2004. Every season DHL has a team of 75 personnel who are tasked with transporting team and broadcast equipment across the world, by air, sea and land. Approximately 1400 tonnes of freight needs to be taken to each event.

The race cars are needed at each grand prix, but these are not simply wheeled into the back of a plane and unloaded at their destination as was the case in the nascent days of the championship. The two cars are stripped down to the chassis and then join the spare chassis in specially designed frames that are secured to a pallet, which is loaded on to the plane. Delicate pieces of bodywork are placed into bespoke foam packaging or bubble wrap that is then boxed, and the shape of these containers must evolve if there are changes to the size of components. Air freight is managed by DHL, and the UK-based teams take equipment to East Midlands Airport, from where six Boeing 777s – which replaced the less efficient 747s – transport the essential freight across the world from race to race. Critical equipment for teams also includes fuel, tyres, engines (and spares) and IT systems.

'F1 will tell us what time our freight needs to be at the airport and then we work back from that point,' explains Crolla. 'We work with our transport company who's picking up the air freight containers to know that, *OK, well if we need to deliver it into East Midlands by 10 on Friday morning, we know that it needs to leave our factory by 5 on Friday morning,* but we will give ourselves some margin as well.'

There have sometimes been complications despite DHL's expertise. Teams have to complete the paperwork meticulously, particularly in countries with more complicated bureaucracy, to ensure that components are not held up in customs. Inflation through 2022 pushed up costs and led to discussions over greater allowances in the cost cap, while the conflict in Ukraine meant Russian airspace was no longer usable, resulting in the lengthening of some flight paths. In 2021 a storm in the Caribbean caused one flight to be diverted, impacting its transit and meaning some teams received their equipment a day late in São Paulo and were given a waiver to the standard curfew that dictates the hours personnel aren't allowed in the paddock overnight.

When freight does arrive as scheduled in a country, teams must still wait.

'As with any air cargo arriving in a country, it has to clear customs,' Crolla explains. 'They'll check obviously that the consignment contains the containers that we say it does and then it will get shipped by F1 cargo again or their local handling agent to the circuit. And then there's a time at which every team is able to access our freight at the same time. So typically that's 8 on a Monday morning wherever we are.

'We've got these huge, white airfreight containers that all our equipment and cars travel in. They're normally unloaded in a specific order, depending on what each one contains, so that you know that garage equipment will be unloaded before cars. And then by the end of it, cars are normally the last thing to come out and then your empty airfreight containers are then closed up, secured and either transported to an on-site storage location or picked up by a freight handler and taken somewhere else.'

Teams also have non-critical equipment that can be stowed in sea freight, which is significantly cheaper, meaning that six identical versions of the same component can be shipped around the world. The sea freight can include items such as garage panelling, tools, the pit wall, hospitality tables and chairs, fridges, cookers, lighting, long-lasting consumables, desks, cables and grid trolleys. This process starts for the next season before the current season has even finished to ensure that the equipment can be sent around to different pockets of the world.

'We know that going into the first race, the sea freight set that we were using for that event had to leave us in December,' says Crolla. 'So you've got to track back, you know, months in advance because of the amount of time it takes to transport that equipment.'

One batch may cover the likes of Australia and Singapore, another may do Azerbaijan and Japan, another can remain in the Gulf region, another may do Miami and Mexico City.

For European grands prix, teams have their own fleet of trucks that transport equipment across the continent, but there are still tight deadlines to meet and logistical challenges, particularly in paddocks with trickier access routes. The racetrack becomes a lorry parking lot within hours of the chequered flag – and pack-up begins during the race itself – and hundreds of boxes festoon the pit lane, with everyone from the team helping – sometimes even the drivers too – to ensure a smooth departure.

'Typically we'll be given a time by Formula One Management as to what freight movements can happen, at what point,' Crolla explains. 'So we can do a certain amount during a race, after the race is finished, though normally there's no freight movements in the paddock until the vast majority of the people have cleared from there, which is always a bit of a battle at certain locations. And then once your main air freight containers are accessible, you'll see each team bringing their forklifts in, bringing their containers into the pit lane and on to the track, and they're loaded in the correct and planned order.'

By the time most non-logistics personnel are trickling out of the paddock, they have to do so weaving in between lorries, the forklift trucks that are zipping around, and the boxes and pallets being loaded with all the paraphernalia.

'There's a lot of equipment that needs loading in a pretty short space of time,' says Crolla. 'Because typically for an average Sunday afternoon race, your air freight has got to be ready by late night, be it 23:00 or midnight, so that it can be transported on to the next event. Now, sometimes we might be in a back-to-back race situation whereby you'll be given an allowance of priority freight. So it might be that you'll designate three or four pallets as the ones that need to be there earlier. And those are the ones that you release first on a Sunday night and those [go] back to F1 cargo and they transport them onwards first so that they're at the next race earliest for you to access them.'

In the case of back-to-back European events, the first trucks will leave on a Sunday evening and there will be multiple drivers on hand for safety purposes and to comply with directives on maximum time behind the wheel. By Monday afternoon the first trucks will begin arriving at the next event and park along the start/finish straight, to begin the unloading process for that week's grand prix. By Wednesday afternoon the trucks will depart the scene to park up in a nearby location – for Monaco this is a lorry park just across the border in France – until they're drafted back into action a few hours after Sunday's race.

It is an arduous job and it has to run like clockwork. After all, nobody really notices the work of each team's logistics people until there is a slip-up and something isn't ready or in the right place.

'When a lot of us first started in Formula 1, we had calendars of 18 or 19 races and now it's grown to 24,' says Crolla. 'The pressure that puts on the human element and the equipment element should not be underestimated. We're really at the tipping point of the sport between being able to function properly, safely, and find people who are still willing to do these roles that require that level of travel.'

There has been increased staff rotation in some departments, usually within marketing and communications, on account of the expanding calendar – such as ensuring personnel do not work all three races in a triple-header – but this strategy has its limitations.

'We can't have a "B crew", because there isn't the money in the budget cap to allow you to do that,' Crolla explains. 'And you also can't be going from one race to another not having that experience transferred from race to race. The guys at one race need to know exactly what happened at the previous race to pick up on care points and the reliability issues.'

Fatigue – some natural and some psychological – also plays a role for team members through the course of a 24-event season.

'You could plot a graph as to when your lows and highs of fatigue are going to be,' Crolla says. 'Those are the races where you know that everything needs to be perfectly executed because people are going to be short-tempered, they're going to be tired, they're going to be really looking inward as to say, *You know, how much longer have we got at this?*'

'People are physically tired when you come to the end of a triple-header, for example. But because people know that there's a summer shutdown coming, you know that everybody was at the end in the last race before the summer shutdown. You know that when they get to Abu Dhabi, everybody's going to be at their lowest level as well. Triple-headers really wipe people out and, having done plenty of them myself, there's the psychology of being away from home and your family for that period of time, but also the physical nature and the psychology of just working from one race to the next race to the next race. It seems like it's never going to end. And by the third race of a triple-header, you're done.'

There is a widespread consensus that 24 grands prix is the limit – and, in fact, many argue that 20 events is the ideal number. But new events are jostling for prime position with existing grands prix in a sport that is striving to find the right balance while chasing the financial injection that comes with expansion.

Miami, among the newest and shiniest of Formula 1's toys, is not part of a triple-header, but it comes at the end of a sequence of six events across two months in three different continents, bringing a gruelling opening phase of the season to a close.

The Hard Rock Stadium, home to the Miami Dolphins,
is transformed into the F1 Village

8

EMILIA ROMAGNA GRAND PRIX: AUTODROMO ENZO E DINO FERRARI, IMOLA

'Imola is such a crazy track and a track I really enjoy to drive on. It's one of those tracks which, I think, now that Formula 1 has evolved, is a bit more difficult for racing and it's a bit more difficult to overtake.'

CHARLES LECLERC

Formula 1's calendar expansion means that the first round in its traditional European heartlands now doesn't take place until mid-May, as the seventh event of the season.

The honour of opening the European season now goes to the grand prix hosted by Emilia Romagna, the culturally and financially rich region of Italy. Imola, a short drive from Bologna, nestles between the verdant Apennines and the agricultural fields, vineyards and orchards that stretch out towards the Adriatic coastline, and the event is the first in a sequence with more straightforward travel arrangements, no jet lag and less time away from home.

Imola first joined Formula 1's calendar in 1980, hosting that year's Italian Grand Prix as a stand-in for Monza, then under renovation. A year later Imola returned to the schedule, but with

Monza back as the Italian round, Imola was instead installed as the San Marino Grand Prix, in deference to the microstate around 100km (60 miles) away.

It is a picturesque and narrow circuit in quiet surroundings, which has been touched by triumphs but also tainted by tragedy.

Imola's grand prix in 1994 altered the landscape for Formula 1. Rookie Roland Ratzenberger suffered a fatal accident during qualifying, and the next day three-time champion Ayrton Senna was killed in an accident during the race.

A single death would have triggered alterations, but the devastating loss of two racers, one of them a global sporting icon, was transformative. There were instant regulation changes to reduce speeds, a trend that continued across the following years and decades, and a renewed push for enhanced safety measures on the cars, including increased cockpit protection and stronger helmets. A widespread analysis of circuit conditions was also undertaken, in particular corners that were deemed to be high-risk, including at Imola.

The high-speed Tamburello curve at which Senna crashed – and where Nelson Piquet, in 1987, and Gerhard Berger, in 1989, had previously suffered sizeable but survivable accidents – was modified into a chicane from 1995, as was the Villeneuve sweep where Ratzenberger died.

It is impossible to disassociate Imola from the grim events of 1994. A statue of Senna, sitting with his head bowed, is often adorned with flowers, hand-written messages and Brazilian flags. Located in the Acque Minerali park behind the circuit's fence, and often shaded by the thick canopy provided by the trees, it is a place of pilgrimage for fans. Tributes are affixed to the fences, flags are attached to the railings, and in 2024 – 30 years after Senna's accident – the current Formula 1 drivers gathered at the statue for a brief ceremony, organised by four-time World Champion Sebastian Vettel, who also demonstrated a 1993-spec McLaren Formula 1 car in which Senna claimed his last career victory.

'I think it is very important,' said Vettel about the tributes. 'One, to understand the risk that drivers are prepared to take at that time, losing their lives doing something they loved, but more so if you look at Ayrton and his story – sure, he had been very successful doing championships and races and some incredible stats that stand to this day – but also the incredible compassion and courage he had to speak his mind. I think it's a very important and powerful story to share, especially with young drivers coming up, or drivers that don't remember or weren't born at the time and will take over in the future in another two, three, five or ten years. [They need] to be inspired by those stories.'

Imola remained on the calendar in its altered form for another decade, but as Formula 1 expanded to new territories in the early 2000s, to the likes of Bahrain and China, its position came under threat. Imola fell by the wayside after 2006, when Michael Schumacher was the victor in a fairly processional encounter, and it appeared improbable that the championship would ever return.

However, the Covid-19 pandemic in 2020 handed Imola an unexpected olive branch.

Formula 1 needed European venues, and ones that were up to the required standards, to fill the gaps in a calendar that had been poleaxed by uncertainty and travel restrictions.

Imola, the Automobile Club d'Italia and the regional government of Emilia Romagna all recognised an opportunity, and secured a spot in the calendar in the autumn of 2020. Then the pandemic continued into 2021, sealing an additional round for the spring. A surprise four-year deal followed and the 2022 event finally took place in front of spectators, after the 2020/21 lockouts – won in 2020 by Lewis Hamilton and 2021/22 by Max Verstappen.

Then, in 2023, came misfortune. Violent storms ravaged Emilia-Romagna throughout May and torrential rain early in the build-up week devastated towns and communities. The weather paralysed the region: lives were lost, and emergency services were inundated with calls to rescue residents and help those stranded. The luscious green fields were submerged and the red-roofed houses poked through the muddy brown water that cascaded through the area. Landslips in the nearby Apennines destroyed roads and, though the circuit itself was spared from the flooding, the adjacent Santerno river overflowed into sections of the paddock.

In the circumstances, Formula 1 had no choice but to cancel the grand prix on the Wednesday before the event and efforts were made to assist with the relief campaign. Some personnel landed at Bologna to news of the cancellation and immediately about-turned to fly back home. Others already on the ground helped where they could, while financial donations were made, and food and water intended for the grand prix was repurposed.

Fortunately Formula 1 was able to return to rustic Imola, and Emilia-Romagna, in 2024 and renew its affiliation with the venue. It is a historic region rich in motor-sport heritage, making its return to 2020s-spec Formula 1 something of an old-fashioned outlier, a situation smoothed by CEO Stefano Domenicali hailing from the town. He volunteered at the circuit as a youngster, studied at the nearby University of Bologna, and spent two decades at Ferrari, rising to the role of Team Principal from 2008 to 2014. After spells at Audi and Lamborghini, Domenicali was appointed as the boss of Formula 1 from 2021.

The event's infrastructure isn't exactly old and creaking, but nor is it new, and it is a compact venue, meaning logistically everything is squeezed into a relatively small area, and some facilities are temporary. This is all part of the quaintness of Imola, a circuit with pastoral houses dotted around the public roads that intertwine the track, meaning those with vantage points cram on to balconies, or the roofs of their houses, to clamber for a prime view of the action. Some will scramble on top of fences, others will perch on temporary scaffolding, and other ingenious solutions have included groups renting cherry pickers to elevate themselves to a higher view. Some climb the trees, or peek through the bushes and thickets that line parts of the fence; others set up shop all day in fold-up chairs beneath the leafy canopies. It feels a little like a town fair, but this is an international sporting event where the cars racing at 320km/h (200mph) are the stars of the show.

'I'm glad the calendar keeps this sort of venue because I think it reminds us all where we all come from and why we all became fans of this sport,' says Carlos Sainz.

It is a narrow circuit – with the best overtaking opportunity along the main pit straight (which has a few curves on it) – while drivers have to attack the kerbs and leave not a millimetre to spare come qualifying.

'Imola is such a crazy track and a track I really enjoy to drive on,' says Charles Leclerc. 'It's one of those tracks which, I think, now that Formula 1 has evolved, is a bit more difficult for racing and it's a bit more difficult to overtake. But it's one of the most exciting tracks for the qualifying lap and this, I think, we all enjoy as drivers.'

Many of those balconies of the apartments behind Rivazza corner, and the grandstands inside the circuit, are festooned with Ferrari flags. The circuit, officially the Autodromo Internazionale Enzo e Dino, is named after Ferrari founder Enzo, and his son Dino, who died at a young age. Scuderia Ferrari is based in Maranello, around an hour away and the other side of Bologna, the large city to the north-west of Imola, famous for its cuisine, porticos and university. You reach Bologna if you leave the circuit and turn left at the main road, but if you instead turn right it'll be only a few minutes until you arrive in pretty Faenza, a town of 70,000 known for its ceramics industry and as the home of Italy's second Formula 1 team. Yes, the red of Ferrari is famous – though more associated with the Italian Grand Prix at Monza – but Imola is a certified home race for the team currently competing as RB.

The Minardi team joined Formula 1 in 1985 and established itself as a popular minnow, occasionally pulling off a heroic underdog result, and it was where the likes of Fernando Alonso and Mark Webber got their first opportunities. When Red Bull acquired the ailing outfit in 2005, part of the deal included retaining its Italian roots and ethos. Toro Rosso – Italian for Red Bull – was therefore born, before being rebranded as AlphaTauri in 2020, to promote Red Bull's burgeoning fashion label. Then from 2024 it became known as the cumbersome Visa Cash App RB.

'It's like a second home grand prix,' says Yuki Tsunoda, who made his debut with the team in 2021 and who has lived in Faenza. 'It's one of the closest tracks from the factory and we normally do an off-season test there. I always stay at my house for the race week, which is just 20 minutes away. The people around there are really passionate about motor sports; they're really kind, respectful and they cheer a lot for every driver. It's quite nice and I feel more fans around that. It's different to other grands prix, it's more like to say thank you, and to appreciate the amount of work the people do throughout the year who work in Faenza. People in the factory usually cannot come to a track to see the performance, so Imola is the opportunity where they can come to the track and see it physically. So for me the grand prix is quite important for us, and also for the team, to perform well in front of them to kind of say thank you.'

Daniel Ricciardo raced for the team in 2012/13 and returned mid 2023. 'It really is like 15 minutes away from here,' he says of the factory. 'So it's special, you know, to have a team so close to obviously a venue we race at. And also Italian; you know, it runs, it definitely runs pretty deep in the blood here in Italy.'

Some personnel, such as RB Head of Sporting Direction Marco Perrone, are lifers, having joined at a young age, and the presence of RB facilitates the achievement of a childhood ambition.

'I always dreamed, since I was a child, to work in Formula 1 and this team gave me the opportunity to make my dream come true,' Perrone says. 'They gave me unbelievable opportunities, starting from the last of the engineers straight from university, with no idea of not just Formula 1, but work in general! Faenza is quite a small town, but as a team we've grown up dramatically and have obviously attracted a lot of people from different countries – we're a small team in a small community, but we are now an international team. At Imola, you feel the passion. You feel the people. A lot of our families and friends come to watch the race. So you really feel that there is an extra passion there, and it's great to have a race 15km [9 miles] from our headquarters.'

The opening round of the European season means it's a return from hibernation for the gargantuan motorhomes.

Formula 1 teams had little need to consider hospitality in the early days – given the slim number of personnel congregating together – but as the championship's popularity grew, team numbers swelled, the sport gradually morphed into a business, and there was a significantly greater importance placed upon looking after personnel, sponsors and guests.

Non-European circuits have small buildings for each team, and for Formula 1, the FIA and tyre supplier Pirelli. There is also a hierarchy: the world champions are usually at the top end of the paddock and the last-placed team at the bottom, separated by a couple of hundred metres.

At European rounds every organisation must bring their own units. Immediately behind the pit lane are trucks that contain two-storey engineering offices, usually where engineers and drivers conduct debriefs on the upper floor, and where the tyres and other paraphernalia are stored below. These are only for team members, the secrecy guarded by translucent windows, and are an area of solitude for throngs of the engineering workforce.

The more heralded structures, which take pride of place in the paddock, are the motorhomes. These started out as rudimentary gazebos but have since grown to resemble large houses that are transported across the continent. They are lined up in the paddock – dressed in team colours and instantly recognisable, though Red Bull's two teams share one unit – and are the home from home for the teams during a European grand prix weekend, providing a sense of familiarity. Personnel could almost sleep in them – but fortunately hotels rooms are more comfortable.

Each team has a different design, the exterior and interior reflecting their identity, with aesthetics and auras that leave a lasting impression. Some motorhomes have multiple levels, including features such as a fancy rooftop terrace and bar, and can be dismantled in one country on a Sunday evening and rebuilt – in identical fashion – in time for the next event just a few days later. There are also special features: Red Bull's enormous hospitality structure for Monaco has a large deck that has several bars, replica Formula 1 cars, a swimming pool and a berth for speedboats out back, and is built up in Italy before being sailed on a barge to take up its position in the harbour. Access is by invitation only.

Each team has a designated area, down to the centimetre, for its motorhome. The various units are then installed in a specific order – which will help when it comes to deconstruction a few days later – and the motorhomes built up around central pillars. Once the key structures are in place the ancillaries can be installed, such as pipework, electric cables, and everything that will be hidden from view once walls and floors are fully fitted.

McLaren uses a two-storey unit known as Team Hub, which in 2021 replaced its Brand Centre, an imposing space-age structure that was introduced in 2007. The equipment and personnel required varies for each motorhome. McLaren needs eight trucks, four for the motorhome and four support trucks, and 12 people take 16 hours to erect the structure, and around the same time to dismantle it. As with anything in Formula 1, the scheduling is tight and the process timetabled to the hour, though there can still be setbacks, particularly with tight turnarounds involving lengthy trips. Accidents, tyre failures on trucks, and delayed or missed ferries can impact the build process, and crews must follow guidelines on health and safety, and time limits, but the motorhomes are always ready, and spic and span, for the start of a grand prix weekend.

'The team hub is very much our "business as usual environment",' says Mark Norris, McLaren's Trackside Operations Director. 'We have two driver rooms downstairs and each has a bathroom in there, so if they come out of the car they can have a shower, get in their team kit before they go into the engineering truck. The truck where we have the driver rooms downstairs, above them are our marketing and logistics offices.'

At the back of the ground floor is McLaren's catering area which, as with its competitors, is a fully fitted commercial grade kitchen. A small electrical fire at the 2024 Spanish Grand Prix caused damage to the motorhome, but swift work and renovations meant the Team Hub was operational again just 10 days later.

'We have a walk-in cold store, we have a pastry section, we have induction hot cooking, we have warm cupboards, we have heat lamps, we have a Bratt Pan,' Norris explains. 'And of course with our catering team, they have got two Michelin stars. We have very high-end cuisine for our partners and our shareholders and our guests, but it's the same kitchen that is producing performance-driven food for our race team.'

Above the kitchen McLaren has another two driver rooms with a sofa, PlayStation, TV and massage bed, allowing space for recovery and sessions with performance coaches, while there is also a boardroom with audio-visual facilities for meetings. On the other side of the second floor are three other office spaces, as well as the private office of CEO Zak Brown, and on both floors are communal areas in the middle for general socialising – almost a sacred place away from the chaos of an event. Each motorhome has similar spaces in order to accommodate the plethora of needs for a grand prix team.

'The way that we operate is that the downstairs floor is very much for the team,' explains Norris. 'We try and make that a home-from-home environment. They feel very comfortable when they come in, and it's also where our media guests come into, and our general guests. Then the top floor is really reserved for our shareholders, our partner guests and our VIPs and their guests, and we try and make that a little bit more luxurious.'

The spotlessly maintained motorhome is the team's life blood through the course of the weekend. It is where drivers and their entourages can relax away from prying eyes, where mechanics and engineers can congregate away from the garage, and where A-list celebrities can quaff champagne and dine on exquisitely crafted meals. The motorhome has become a symbol for Formula 1's alluring excesses and extravagances. And straight after Imola, pack-up will begin for the motorhomes to make the relatively short sprint – by Formula 1 standards – to one of the championship's most famous events.

Imola winds its way through the scenic Italian countryside

9

MONACO GRAND PRIX: CIRCUIT DE MONACO, MONTE CARLO

> *'You also don't really have time to think: there's no real long straights; there's no fresh air coming through because it's so tight and twisty through the streets. So, it's all like you're just in it and you can't really get out of it.'*
>
> DANIEL RICCIARDO

It is a place where you can be divided between the haves and the have yachts, where whatever credit card you have probably isn't sufficient for the boutiques and where the racetrack becomes a dance floor every night: Monaco.

Monaco and its grand prix are unmistakably intertwined.

Any mention of the petit principality, perched on a handful of rocks between France and the Mediterranean, immediately evokes thoughts of glitz and glamour, the casino and, most prominently, motor racing. Monaco's grand prix was first held in 1929 on a circuit that still resembles the current layout and in 1950 was part of the inaugural Formula 1 World Championship season. Only Monza has been graced more times by Formula 1 than Monaco's narrow and winding streets. The event, once always linked with the Ascension Day holiday, is now affixed to late May.

It is a remarkable undertaking: a major annual sporting event in the world's second-smallest nation, which has an area of just two square kilometres and can be traversed on a brisk walk. There is such little breathing space left that land reclamation projects to pack in more high-rise apartment blocks where seven figures are required for a broom cupboard are near-constant.

The grand prix is the best of events and the worst of events. It is astonishing yet boring, a total thrill but also a headache at which to work. It is anachronistic yet remains essential. It is a staggering circuit but dreadful for racing. It is an event of contradictions.

Organising it takes metronomic planning and is a laudable achievement, yet security and police are regularly aggressive and unhelpful. Friday's access roads become off limits on Saturday but are in use again on Sunday, while police shrug when politely asked for help and shout angrily at motorists trying to follow rules. Access by rail is possible and advisable for spectators, though there is no incentive for France's rail company to enhance the service for Monaco's grand prix. Moped riders also zip around the principality, ducking and weaving in between motorists, using any spare patch of tarmac or pavement.

Monaco used to be Formula 1's unparalleled blue riband event, the grand prix swooned at and swooped upon by celebrities, VIPs and sports stars. It was where the bulk of business deals were carried out and Monaco itself had hugely favourable promotional terms. The race had its own special timetable, with practice shifted to Thursday, and Friday a rest/recovery day, when Formula 2 took pride of place on the schedule, before action resumed on Saturday. In 2022 Monaco fell into line with the other events, with a three-day schedule and, from 2023, a revised contract meant it lost special financial privileges as well as the control of the world feed. The addition of other high-profile city-based events, such as Singapore and Las Vegas, has diminished Monaco's exclusive aura, while the quality of racing has long been poor, owing to its narrow streets, a situation accentuated by Formula 1's longer and wider cars. Designs were tentatively drawn up to extend the circuit, potentially creating an additional overtaking opportunity, but these were swiftly dismissed by the Automobile Club de Monaco, an organisation that tends to block its ears to constructive criticism. Monaco's promoter believes it has the best organisation. Those with experience of the other 23 grands prix believe otherwise.

Monaco would never be added to the calendar in a modern era. The circuit would never get the required sign-off, the logistics would be deemed illogical to the extent of insurmountable and the disruption would be regarded as too significant. It is a racetrack where racing is borderline impossible.

Yet it remains an absolutely spectacular event. And because the circuit – the majority of the layout is traceable to the one first used in 1929 – is enthralling, but racing problematic, it means Saturday is usually a more gripping display compared to Sunday, creating an atypical weekend for the field of 20.

At most grands prix, practice sessions are usually split between preparation work on low fuel and then longer runs on heavier fuel loads – often referred to as race simulations. Drivers and teams may also have new specifications of updates on various parts of the car to evaluate, alternative set-ups may be trialled to extract maximum performance, and sometimes new systems and protocols may be assessed. But in Monaco there is a heavy focus towards just the single-lap pace. That's because Sunday's race is as close to a procession as possible in motor sport, with 60 of Monaco's 70 races won from the top three starting positions. Overtaking is not completely impossible, with a slim chance of a move into the first corner or the chicane after the tunnel, but it is highly improbable due to the relatively short straights and the barrier-lined narrow track layout, meaning having a pace advantage is meaningless if a slower rival has track position. Teams have to be on the ball when it comes to strategy and track position, but the easiest way of securing the latter is to nail the single-lap pace on Saturday afternoon and then dictate the race pace. Drivers can lap 10 seconds slower in the race than in qualifying and still comfortably hold track position.

That means using the three practice sessions to get a car into its optimum window for performance over a single lap come Saturday afternoon, while adapting to the evolving track conditions as rubber gradually gets laid down. For the drivers the running is vital to put it all on the table for a 70-second blast around the streets, surviving the risk of being eliminated in Q1 and Q2 and then shooting for pole in Q3.

'Saturday, Monaco, it's probably the biggest day in F1 over the year, and it's definitely the biggest day if you get it right,' says Daniel Ricciardo, a two-time Monaco pole-winner. 'That feeling of putting it together. Every circuit requires full commitment, it's obvious, but this is another . . . there's more variables, there's more factors, there's just something else. And you also don't really have time to think: there's no real long straights; there's no fresh air coming through because it's so tight and twisty through the streets. So, it's all like you're just in it and you can't really get out of it, if you know what I mean. So when you hook the lap up, it's . . . yeah, you feed off each corner: every corner you do well, you link the next one up and it's a pretty beautiful feeling.'

Drivers also have to ramp up their preparation through the weekend and not run before they can walk.

'I think the main approach is you always have to start under the limit and then try and find the limit because there's not a lot of room for error, but at certain circuits you can easily go over the limit,' says Lando Norris on the challenge of street circuits. 'Places like Monaco, Singapore, Suzuka in a way, you can never go over the limit because if you go over the limit it'll bite you pretty hard. So you always have to start under it, which is a much better challenge, I think, for drivers. I think the mentality needed for that is quite a tricky one. And yeah, it just puts you mentally in quite a weird place because you just feel like you're risking so much half the time. It's just a scary place to be. It's just like a cool feeling you get, it's very unique and special of, like, *I feel like I'm about to crash*. But you don't.'

Says Fernando Alonso: 'It's very unique and probably one of the best moments of the season. I agree on that. The whole weekend, I think, is very special. Maybe only Sunday is the bad day, let's call it; you know, it's not much you can do. It can be a little bit boring as well. You just need to bring the car home and things like that. But until Sunday, I think the free practices are very interesting. The circuit is evolving at a rate that is not [the case] in any other circuits. So you have to guess, you know, what the grip will be in the next session. All the set-up changes, they have implications of guessing what the car and the track will do. And then in qualifying, [it's] the only qualifying of the season that basically you go through corners at the speed that you've never been before.'

The circuit's corner names also reveal some of Monaco's landmarks. The first corner, Sainte-Dévote, is a nod to the adjacent tiny chapel, constructed in honour of Monaco's patron saint. The two Mirabeau curves are named after the old hotel nearby; Tabac takes its name from a tobacconist's on the outside of the corner; the fearsome Piscine chicane goes around a public swimming pool; and there's also Nouvelle Chicane, the 'new chicane', a new addition for safety purposes back in, er, 1986. The circuit also passes some of Monaco's most iconic buildings, such as the Hotel de Paris, where its outdoor patio rubs up against the crash barriers, as well as the Casino de Monte-Carlo, the landmark which transformed Monaco's fortunes and where Monégasques are still not permitted to gamble.

Monaco is not exactly renowned for its sporting landscape, beyond its grand prix, historic rally and Ligue 1 football club – which plays in the Stade Louis II in Fontvieille and can't be said to have a massive following – but it is home to a plethora of sports stars, including a sizeable portion of the Formula 1 grid, some even living in the same block of apartments. They are enticed to Monaco by its convenient location, all-year-round good weather for training and lifestyle, as well as the exclusivity and privacy. The amenable tax rules are also a sizeable tick in the right box.

But Monaco has one of its own in the form of Ferrari driver Charles Leclerc, a bone fide Monégasque in the only territory in the world where nationals are outnumbered by foreigners. Leclerc was born and bred in Monaco, his first memory of Formula 1 is watching the race from a friend's balcony, and he went to school on the rock of Monaco-Ville, located above the paddock. This is the homeliest home race possible.

'For me it is weird because it's the city which I grew up in and to see the city just changing completely for the Grand Prix is very, very special,' said Leclerc. 'All of my friends, all of my family is here, watching from balconies of apartments of my friends around the track. To me it feels a bit like a village, but obviously on an event like [the grand prix] it's a bit bigger than a village and a lot of people are coming to Monaco to watch the race, so it feels special.'

Leclerc is a rarity in being an actual Monégasque – rather than someone who relocated – and his career at home grands prix did not start well.

In 2018 Leclerc suffered a brake failure, which cannoned his car into a rival, and in 2019 a Ferrari blunder in qualifying left him mired down the grid, from where he was overly aggressive and suffered car damage in the race. He claimed pole position in 2021 and 2022, but still this led to heartbreak: in 2021 a crash at the end of qualifying – which ironically secured him pole position – caused car damage and though Ferrari extensively checked the car, a driveshaft on the undamaged side failed en route to the grid. In a rain-hit 2022 race, Ferrari dropped the ball strategically and he dropped to fourth. Leclerc qualified third in 2023, but a penalty dropped him to sixth, meaning even a podium finish remained elusive.

Finally, in 2024, Leclerc was the man all weekend, flying out of the box from the start of Friday practice, and he converted pole into an emotional win. It was a victory received well by the entire paddock and by his contemporaries. Even marshals trackside were in tears and Prince Albert II also shed a tear on the podium; the two first met when Leclerc was a teenager seeking support for his career.

'Monaco is the grand prix that made me dream of becoming a Formula 1 driver,' Leclerc said after his victory, before diving into the harbour. 'Yeah, I remember being so young and watching the race with my friends, obviously with my father, who has done absolutely everything for me to get to where I am today, and I feel like I don't only accomplish a dream of mine, but also one of his. To finally make it in front of my whole family, my friends that were watching all over the track is a very, very, very special feeling. I realised actually two laps to the end that I was struggling to see out of the tunnel just because I had tears in my eyes. And I was like, *Fuck, Charles, you cannot do that now. You still have two laps to finish.* And especially on a track like Monaco, you have to be on it all the way to the end. It was very difficult to contain those emotions, those thoughts again, of the people that have helped me to get to where I am today. It's only one win. The season is still very, very long. It's 25 points like any other win. However, emotionally, this one means so much.'

Leclerc may have termed Monaco a village, but it does not exactly have open green spaces, chocolate box cottages and a cricket pavilion. Monaco's compact streets are dripping in wealth – the circuit passes in front of upmarket boutiques such as Hermès and Gucci – and obscene displays of money are rarely far from view. The yachts in the harbour – the prime berths command upwards of six figures for the week – are complemented by larger boats and cruise vessels that rest in the glistening azure sea of the Mediterranean, sometimes accessible by speedboat transfers, and there is a regular hum of helicopters whizzing from the heliports of Monaco to Nice Airport, a journey that takes seven minutes and costs a couple of hundred euros. It is essentially a race in a tax haven, a rich paradise with gourmet food and free-flowing champagne (though the workers improvise with a homemade three-course meal for lunch, frying sausages on disposable barbecues, cooking paella trackside, and ensuring there's a plentiful supply of baguettes). The rocky backdrop is spectacular too, though the scenery is just as gorgeous along the coast, to the west in France or to the east towards the town of Menton and Italy. Local radio stations are not festooned with payday loans adverts but rather promote asset management companies, yacht insurance and supercar showrooms. The press room is located inside a long terracotta building that also houses yacht storage and servicing units, an international school and, until its closure in 2022, a favourite haunt of personnel: the Stars 'N' Bars diner.

The restaurant had been frequented by the great and the good of the sporting world, with more memorabilia than most museums, and its centrepiece was a full-sized 1999 McLaren F1 car, affixed to one of the walls. For years it was also the place at which to watch the Indianapolis 500 – which regularly takes place on the same weekend, on Sunday evening – the majority present tending to cheer on the ex-F1 contingent now racing in America.

At the top end of the paddock is the famous Rascasse restaurant, which effectively acts as the apex of one corner and which turns into a nightclub every evening. The racetrack becomes the dancefloor for several outlets, blasting out a range of music, while in the pit garages above mechanics continue to fettle the cars until curfew. Opposite Rascasse is one section of Monaco's famous harbour, where the yachts and superyachts arrive for the week, with the most sought-after berths requiring a six-figure sum. They are squeezed in alongside one another, lining the perimeter of the circuit, though for safety reasons they must be moved back a few metres during track activity. The parties onboard continue through day and night, with the end of the race marked by a celebratory cacophony of foghorns. It is the epitome of decadence, albeit requiring the spectator to overlook the squawking seagulls and the intermittent stench of sewage.

It is the most challenging and claustrophobic working environments of the year for paddock personnel. It has been improved over the years, but the garages for mechanics are cramped and set across multiple levels; some even have trees and branches sticking through them. The team 'pit walls' are relocated to one of these floors and, since a restructure in 2004, the pit lane is the only one on the calendar that doesn't overlook the start/finish straight, while the pre-race grid is congested with people to the extent that it becomes jammed. The usually gargantuan motorhomes are dwarfed by the surrounding buildings and topography, perched in a line along the harbour, and the paddock is just a thin walkway between the motorhomes and the water.

All this adds to the unique logistical challenge of Monaco: different team departments are splintered across the area and additional time has to be factored in due to the bottlenecks that build up at pinch points. Still, it is at least more convenient than Formula 1's support series: Formula 2 sets up its paddock on the ground and first floor of a multistorey car park carved into a rock around the corner, while Formula 3's paddock is not even in the same country, as it has to relocate across the border to France.

Monaco is one of the best venues all year for getting up close and personal with Formula 1 cars. There are some sections of track where it's possible to perch safely just inches away from the racing line, and where observers get slapped by the extreme heat from the cars and spattered by the tyre marbles and assorted debris flicked into the air. Stand at some turns and it's as if the drivers are aiming directly at you before darting elsewhere at the last possible second. It is the one circuit where the speed of the cars is just mind-boggling and the laws of physics seem like they are being scrambled, and where it is scarcely believable – and a testament to the supreme skill of the drivers – that there are not more accidents. Monaco's ultra-responsive marshals are also swift to caution anyone standing in a position that could be too perilous. You can't react fast enough to a Formula 1 accident in close quarters – it will happen before you know it.

Standing trackside throughout the course of the season, and capturing those mesmerising on-track moments, is the job of a photographer. It is their imagery which will be used by newspapers, magazines, websites, billboards, promoters, sponsors and more, freezing moments in Formula 1 history forever.

'Like everything in F1 it is really intense,' says Florent Gooden, a photographer for French agency DPPI. 'Even though you have two or three hours of cars running during the day, basically it doesn't stop there. You have a lot of off-track stuff to shoot, such as driver portraits, pit lane stuff, marketing activities, there'll be VIPs and guests, you want to get the atmosphere of the place, or the paddock, and sometimes shoot Formula 2 and 3 as well. And in between all this you also have to send the pictures as quickly as possible, so every time you have 10–15 minutes without shooting, you rush back to the media centre to send the pictures back to base before going out again. You have to think about where you're going to stand for every session. For example, during free practice we can go quite far away, maybe on the other side of the track, and we have to keep some places in mind for qualifying or the race that may be closer to the pit lane so we can come back for celebrations and stuff like that.'

Photographers must always wear a tabard when trackside and adhere to protocols, such as not standing in the red zones, which are designated dangerous by the FIA. Choosing where to stand for a session can also change depending on elements such as circuit location: street circuits are typically harder to navigate, compared to permanent venues that will have a circuit perimeter road and frequent shuttles. The weather, angle of the sun and consequently light will also have a major influence, particularly at venues where track action takes place at dusk – a favourite of the photographers due to the stunning orange light – or at night.

'The free practices are always a good moment to take some risks with settings to try some difficult pictures, that are a bit more technical, or a little bit arty,' says Gooden. 'Then during qualifying and the race you have to think more on the journalistic part: getting cars battling, side by side, the celebrations. You have to show what's happening at that event. My favourite track is Singapore. There is always something new to try, as one difficult part of the job is to stay creative. Silverstone or Austria, for example, I've been there so many times it's hard to find something new and Silverstone is not the best for photos anyway. Then Monaco is still Monaco. I think everything has been done around that track actually, but you really feel the speed with the cars going just right next to you while you are shooting. They are also nice pictures there, they're iconic shots.'

Photographers have an abundance of kit that is taken to every grand prix – as much as possible in hand luggage and the rest at the mercy of airlines in hold luggage – and they traipse around circuits weighed down by that equipment.

'You always have to have two cameras, because you need a long lens on a camera and a shorter lens on the other,' Gooden explains. 'If something happens, or if you need to change quickly, you don't have to change the lens itself and potentially miss out on taking the pictures during that short time. So we have two cameras and usually many lenses; I have five in my bag. Usually the long lens is 500mm, in order to shoot the cars from quite far away, and it's the one you use most of the time – even in the paddock, taking portraits of the drivers or other people.'

Being a photographer also means being in the thick of some iconic moments in Formula 1 history and capturing those moments for future generations.

'We tend to forget this because we are so focused on the job at the moment, that these pictures will be seen maybe in 10, 20, 30 years,' Gooden says. 'When 2021 happened, that crazy, crazy season, and we came to Abu Dhabi and it was just an epic final, it was really a historical moment. And to be there and be able to capture these moments, you know, *parc fermé* and the celebration of Max Verstappen, you know that these pictures, they will stay for a long time. I know that these pictures of key moments in Formula 1, even if I'm not there in 20, 30, 40 years and that the agency is still there, people will go back in the archives and see these pictures. It's exactly how I'm going in the archives and looking at the pictures. I'm like, *Wow, it was like this, back in the day*, seeing a picture of Senna and thinking that a colleague shot that.'

Monaco is an event of contradictions: enjoyable yet frustrating, where the best of Formula 1 is on display in qualifying, while the races regularly produce a soporific Sunday that provokes discussions over its placement on the modern calendar. It is an argument that rages in the aftermath of the race and is then forgotten about for the following 51 weeks.

Would Formula 1 survive without Monaco? Arguably, yes. Formula 1 is now more important for Monaco than Monaco is for Formula 1. But Monaco is one of the few locations that provides a direct line from the early days of motor racing through to the current era as the event creeps towards its centenary. Granted, Sunday is often a spectacle that proves uninspiring, but to weave Formula 1 machinery through the streets for 78 laps remains a challenge and there is a tacit acceptance that Monaco is part of the product. Perhaps there could be circuit tweaks and perhaps Formula 1 could consider format alterations; a lack of evolution risks stagnation. But Monaco – particularly on a Saturday afternoon – remains a wildcard sporting event that transcends just Formula 1.

10

SPANISH GRAND PRIX: CIRCUIT DE BARCELONA-CATALUNYA, BARCELONA

> 'The Spanish GP is one of the biggest sporting events in the country, if not the biggest. . . . It's interesting that the Spanish GP is one of the oldest and longest considering there hasn't been a Spanish team or even a big automotive manufacturer in Spain – only SEAT – and not even a champion until 2005.'
>
> JESÚS BALSEIRO, FORMULA 1 CORRESPONDENT FOR *DIARIO AS*

Spain's grand prix has a familiar home, camped on the outskirts of Catalonia's charming capital, but it is poised to shift location in the very near future.

Spain's first world championship event took place in Barcelona, the upmarket Pedralbes district hosting two street races, in 1951 and 1954. It was over another decade before Spain rejuvenated its motor-sport scene by constructing the permanent Jarama Circuit, on the outskirts of Madrid, a venue that alternated host duties with Barcelona's Montjuïc Park. The Catalan track was a daunting blast through the public roads of Barcelona's beautiful Montjuïc, around the Palau Nacional, with its hilltop location presenting gorgeous views of Barcelona's perfectly patterned avenues and its imposing Sagrada Família.

But neither venue had a truly positive legacy. Jarama was narrow, overtaking was hard and the event was sparsely attended, while Montjuïc's safety standards were queried by drivers. Their foresight was confirmed in grisly fashion in 1975, at the culmination of a weekend in which drivers had threatened not to race: the Hill GH1 of Rolf Stommelen – a solid grand prix racer who was significantly more successful in sportscars – suffered a rear wing failure that sent his car over the barriers. Five spectators were killed and Formula 1 never returned to the venue.

Southern city Jerez, best known for its sherry industry, took over the mantle for Spain's grand prix at a purpose-built circuit in 1986. It held the national grand prix for only five years: the circuit was unremarkable and it replicated Jarama's problem of low crowds. Jerez twice had a reprieve as the host of the European Grand Prix, in 1994 and 1997, its last appearance on Formula 1's calendar being famous as that year's controversial title decider. Three drivers set an identical time for pole position, while championship contenders Jacques Villeneuve and Michael Schumacher collided during the race. Villeneuve continued and won the title, while Schumacher retired on the spot and was judged to have caused the clash. Jerez became a popular testing venue during the era of unlimited running and hosted official preseason testing as recently as 2015.

Barcelona returned in 1991 and did so at a newly constructed permanent facility, which was inaugurated as part of the city's renovation and rejuvenation in preparation for hosting the 1992 Summer Olympics.

'The Spanish GP is one of the biggest sporting events in the country, if not the biggest, with Vuelta a España and only a few more international events that take place in Spain,' explains Jesús Balseiro, Formula 1 correspondent for Spain's sport newspaper *Diario AS*. 'We don't have a Grand Slam in tennis and apart from football and maybe cycling there is not a very long tradition at any particular sport. It's interesting that the Spanish GP is one of the oldest and longest considering there hasn't been a Spanish team or even a big automotive manufacturer in Spain – only SEAT – and not even a champion until 2005.'

Motorcycle competition – the country is one of MotoGP's cornerstones – rallying and other off-road competitions have long been popular in Spain, but Formula 1 boomed at the turn of the century.

The country had produced only a smattering of drivers in its history, the most promising being Alfonso de Portago, but early in his Ferrari career he was killed in an accident at the Mille Miglia road race, in 1957. There were also the likes of Adrián Campos, Luis Peréz-Sala and Pedro de la Rosa, but in the late 1990s there was a promising youngster tearing up the junior categories. By 2001, and while still a teenager, Fernando Alonso was thrust on to the scene by Minardi and performed wonders with a car that was among the slowest on the grid. After a year as Renault's test driver in 2002, Alonso returned to the grid in 2003.

Attendance at Barcelona's race swelled, the blue-and-yellow flags of Alonso's home province of Asturias fluttering high from the grandstands and coincidentally mirroring the livery of his then Renault team. Alonso became Spain's first grand prix winner in 2003 and two years later was crowned Formula 1's youngest World Champion, aged 24, a feat eclipsed only by Sebastian Vettel and Lewis Hamilton. Finally, in his second title-winning campaign in 2006, Alonso won the Spanish Grand Prix and added a second win at the track in 2013. Alonso's popularity in Spain was such that it facilitated the creation of a second race event in the country, which took on the European Grand Prix tag, and which took place around the streets of Valencia from 2008. But the circuit was not popular, a dynamic accentuated by its location in the uninspiring port area of Spain's third city, and the funding was also an issue. It nonetheless had a fitting farewell in 2012, when Alonso was triumphant after surging forward from a poor qualifying result. Alonso started only 11th on the grid but carved his way past rivals, benefited when a couple of drivers ahead retired and celebrated with the marshals after climbing from his car on the cooldown lap.

'The growth of F1 in Spain with Fernando is simply massive,' explains Balseiro. 'It became the second sport in terms of TV audiences and still is, even 10 years after Fernando's last win. He's right there with Rafa Nadal as one of the two biggest sports stars in Spain of the last two decades. The length of his career has also made it possible that even the teenagers nowadays support him, even if they were kids or not born when Fernando won the world championship.'

Alonso has participated in more Formula 1 seasons than any other driver – over 20 – starting as a teenager and continuing into his 40s, and is likely to eventually conclude his career with more than 400 grands prix under his belt. The second half of his Formula 1 career has been less successful – prompting Alonso to show his all-round ability by tackling the likes of the Indianapolis 500, the Le Mans 24 Hours and the Dakar Rally – but he remains a sporting legend, particularly in Spain.

'In the Spanish culture, in terms of sport, the winner is bigger than the sport itself,' Balseiro says. 'There are a few examples of Spanish athletes who became champions in a sport that wasn't big before and Formula 1 is just another case with Fernando. I don't feel a big motor-sport tradition and F1 fan base, but a huge fan base around Fernando. And Carlos Sainz has become more and more relevant in that scenario. He even carried the flag alone for two years with decent results and coverage.'

The Spanish Grand Prix is a special event for veteran racer Alonso.

'Over the years there have been some exciting battles on track and one that stands out is my win in 2006,' he says. 'It was my first victory at my home race and I will always remember the fans' reactions. It was very special. There is always a different kind of energy at the Spanish Grand Prix for me – the fans are so passionate. Another good memory of mine was the blue wave in 2005 and 2006, which is now becoming a green wave.'

Barcelona may evoke thoughts of the artistry of Gaudí, the high-end fashion boutiques and tapas bars selling Catalan delicacies, the boulevard of Las Ramblas and its neighbouring alleyways, and the miles of beaches beneath azure skies. However, Barcelona's Formula 1 track is surrounded by an enormous industrial estate, including row upon row of factories and warehouses, and petty theft is commonplace. Access to the venue has sometimes been a problem: the industrial estate's roads are known to clog in spite of its close proximity to the highway that connects Barcelona with France, and the small railway line behind the main grandstand does not have a station. The closest, Montmeló, is an unattractive half-hour walk away.

'Maybe the thing that Barcelona lacked is a bigger relation between the city and the F1 race,' Balseiro says. 'For too many years, the city of Barcelona has been facing backwards to the race, maybe because of the political situation in Catalonia or the lack of interest in motor sport from the regional government. When you travel to most of the F1 races, you normally feel the F1 atmosphere from the city itself, you even see big F1 banners the moment you leave the plane at the airport, but this wasn't the case in Barcelona. It was corrected in the most recent years, the presence of F1 was becoming bigger in the city during the race weekend.'

The Circuit de Barcelona-Catalunya is known as a test of a car's aerodynamic prowess: the typical sentiment being that if a car works at this track, it will work anywhere. Only five times since Barcelona joined Formula 1's schedule in 1991 has the race not been won by that year's constructors' champions, and two-thirds of its races have been won from pole position – making it frequently a processional affair. Barcelona features a mixture of corners, though as Formula 1's aerodynamics have developed it has become more of a high-speed venue and turns that once required a lift are now easily run flat-out.

'It's a track where all the teams and drivers have done a million laps, so we're pretty experienced around here,' says Alex Albon.

And, despite its reputation, it's a track that has produced some standout moments.

In 1996, a rare wet day soaked the circuit and provided an opportunity for Michael Schumacher to shine. In a difficult first season with cumbersome Ferrari machinery, he recovered from a clutch problem at the start to take the lead and dominated, cruising to victory by 45 seconds in conditions more suitable for boats.

In 2012 the unheralded Pastor Maldonado converted a surprise pole position into an unlikely victory for Williams, on a day in which all the stars aligned for driver, car and team. There was also a fortunate escape post-race when a fire engulfed the pit garage of the victorious Williams team and Maldonado carried his cousin to safety.

In 2016 Mercedes drivers Lewis Hamilton and Nico Rosberg spectacularly collided on the opening lap, and their already frosty relationship reached new icy depths, paving the way for Max Verstappen to capture a shock maiden win on his first start for Red Bull Racing. Aged 18 years and 228 days, Verstappen smashed the record to become Formula 1's youngest race winner, a benchmark that is unlikely to be lowered.

For several years Barcelona was the home of preseason testing, while it is often used for private running in older machinery for young drivers, courtesy of its convenient location and relatively stable weather. It is also a place targeted by teams for introducing upgrade packages, due to its status as a known quantity, though updates now tend to filter through with greater regularity. But Barcelona's days as a test bed of a thoroughbred are already dwindling and are soon likely to be completely numbered, because after three decades Formula 1 is on the move.

In 2026, Formula 1 will relocate its Spanish Grand Prix to a new track on the edge of the capital, Madrid, with the IFEMA exhibition centre the focal point of the venue. At 5.4km (3.3 miles) long, the circuit will incorporate both street and non-street sections across the Valdebebas district, bisecting a highway adjacent to the training ground of Real Madrid. Organisers hope to tap into convenient transport links: a metro line connecting IFEMA with central Madrid and Barajas Airport only five minutes away. The lengthy contract, from 2026 to 2035 inclusive, was announced in early 2024.

'We know that F1 is more than a race, it is an unprecedented opportunity to drive the transformation that Madrid is undergoing and to show the world what we are capable of,' said Madrid's mayor, José Luis Martínez-Almeida, when announcing the deal. 'I am confident that Madrid will be up to the task, not only because we deserve a spectacle of the magnitude of F1, but also because F1 deserves a city with the energy, character and passion of Madrid.'

The long-term contract allows for both Formula 1 and Madrid to plan well into the 2030s, with the aim of welcoming 140,000 spectators daily by the turn of the decade.

'There are reasons to think it will be successful,' says Balseiro. 'It will be the only European race that will take place in a capital city. It will try to look like the modern events like Miami, Vegas or Singapore and it will emphasise the big cultural position of Madrid as a meeting point between Latin and European worlds. This way, Spain will secure a place in the calendar in the long term and Madrid will have a big international event that has been lacking so far. The city tried to organise the Olympic Games in 2012 and 2020 unsuccessfully, but F1 is the event that Madrid deserves.'

For Formula 1 it's set to be *Adieu, Barcelona* and *Holá, Madrid*. And the evergreen Alonso, contracted to Aston Martin through 2026, is anticipated to be on the grid for Madrid's debut, 25 years after his own bow.

Barcelona has hosted Spain's race since 1991, but Madrid holds a contract from 2026

11

CANADIAN GRAND PRIX: CIRCUIT GILLES VILLENEUVE, MONTREAL

'The track is very unique because you actually get to ride some old-school kerbs and the scenery is pretty cool too.'
MAX VERSTAPPEN

Formula 1's Canadian Grand Prix takes place at a park-based circuit named after one of the sport's lost legends, a stone's throw away from a vibrant and popular city.

A quick transatlantic hop breaks up a prolonged spell in Formula 1's European core. The biggest city in Canada's Quebec province, Montreal, truly embraces the sport and turns its centre into a weekend-long festival that signals the start of its summer, having escaped from the shackles of its lengthy and chilly winter.

Canada first joined Formula 1's calendar in 1967 and the event initially rotated between permanent racetracks: Mont-Tremblant in upstate Quebec and Ontario's Motorsport Park. But it was not a long-term solution.

'In 1977 the FIA said to the organisation back then, "We don't want to keep switching, find one track,"' explains long-time Canadian Grand Prix promoter François Dumontier, who stepped down in 2024, 'So they went to Toronto and the mayor back then said: "You know what, I don't think Formula 1's got a future, so we're going to decline."'

Toronto's rejection of the rights to Canada's grand prix meant attention turned to Montreal.

'The organisers were around Montreal for a few years,' continues Dumontier. 'They knew Montreal, they went to see our mayor at the time, Jean Drapeau, after the Summer Olympics we had in Montreal [in 1976]. The first location the mayor proposed was the Olympic Stadium, which is on the edge of the city, because his mindset is: "We've got 60,000 seats in the stadium, we can have the start/finish line in the middle of it," but the FIA were not sure. The mayor said: "I have a second place for you," because he was trying to reuse the Olympic venues and we've got the rowing basin, so this is when he said: "OK, let's go there," and this is how we ended up doing the track there.'

That rowing basin, used for the 1976 Summer Olympics, was created inside the artificial Île Notre-Dame, an island in the Saint Lawrence River. Adjacent to the natural Saint Helen's Island, it makes up the Parc Jean-Drapeau, named after the mayor. Drapeau initiated Expo 67, for which Île Notre-Dame was constructed – using the rock that had been excavated for the city's metro. Expo 67 coincided with Canada's centenary and afterwards the island became a parkland, with trees and blossom, and the rowing basin was also constructed.

A temporary street-style circuit, with medium-speed chicanes, hairpins and walls close to the track, was mapped out around the roads of the island, and Formula 1 moved there in 1978. The inaugural race was won by Canadian icon Gilles Villeneuve, after whom the circuit was renamed in 1982, following his untimely death in an accident at Zolder, in Belgium. The *Salut Gilles* signature, in his honour, remains inscribed across the start/finish line at the circuit.

Villeneuve's son, Jacques, emerged as Canada's most successful Formula 1 racer, winning 11 grands prix and being crowned World Champion in 1997 – the most recent non-European to win the title. Canada has been represented on the grid since 2017 by Aston Martin racer Lance Stroll, and this is a popular venue in the sport.

'It's one of the most exciting weekends of the year in Montreal,' says Stroll, who scored his maiden Formula 1 points at the event in his rookie season. 'It's my home town, it's where I grew up, 20 minutes from the track, I grew up going to the race as a kid when I was six or seven years old, and it was a big part of my year. I looked forward to that weekend all year.'

Downtown Montreal, a short drive or metro ride from Île Notre-Dame, comes alive during the race weekend. As it emerges from its winter hibernation, there are a host of events and activities throughout the week, and the restaurants are packed. (Stroll recommends Lester's for its Montreal speciality smoked meat sandwiches, while the increasingly popular poutine is readily found in the city).

'It's a track we all enjoy going to, one of the more fun tracks to drive if you ask me, and I think a lot would agree, and it's a place the teams love going to,' says Stroll. 'It always gives us some exciting racing, you can overtake there, there's always stuff happening on Sunday. There's also an incredible fan base that loves Formula 1 in Canada and is really passionate about it, it's what makes it a special weekend.'

The wall-lined circuit is deceptively tricky: there are full-throttle blasts linked by a sequence of chicanes and a couple of hairpins at either end of the track which place a particular strain on the car's braking system.

'It's always nice to go back to Montreal, it's a great city and the fans are great there,' says Max Verstappen. 'The track is very unique because you actually get to ride some old-school kerbs and the scenery is pretty cool too. The car set-up has to be a careful balance between straight line speed and being able to run on the kerbs well. We have to find a good trade-off.'

It is a circuit that weathers more than most: Montreal encounters hot summers and brutally freezing winters, which means the city and its roads are in a constant state of repair.

'I personally like most of the older circuits, those that have a bit more character and aren't as flat and wide as many of the newer tracks,' says Kevin Magnussen, a nine-season veteran who has driven for McLaren, Renault and Haas. 'The Circuit Gilles Villeneuve is definitely one of those and it's a track that is very good for racing with the long straight. There are always good opportunities for overtaking and it's never a dull race.'

It certainly wasn't dull in 2007, when Lewis Hamilton converted a maiden pole position into his first grand prix win in a race punctuated by accidents and incidents.

'It's very special coming back to Montreal,' he says. 'It's great to see the people, the city, it's great to see the energy throughout the city as you're going through. I remember my first time here and my first Grand Prix win here, my first pole, in 2007, so it's always special when I come back here.'

Hamilton is one of several drivers to have claimed their maiden victories in Montreal, with Robert Kubica taking his sole win in 2008, before Daniel Ricciardo added his name to the history books in 2014.

Canada has had a June date since 1982, having moved from the autumn, when the weather was too cold and uncomfortable for an international sporting event to thrive. Formula 1 has increasingly striven to regionalise its schedule, for both sustainability and wellbeing purposes, and shifting Canada has been one target. With autumn out of the question, the most logical move would be for Canada to be paired with Miami, but even in May there is the risk that Montreal's weather remains too cold. By the time June arrives conditions have improved, even if there is still a high chance of sporadic chilly and wet days that lead to personnel digging out the winter coats and beanie hats.

'We need the lead-up time of about three months,' says promoter Dumontier. 'Montreal in the winter . . . imagine -20°C [-4°F] or −25°C [-13°F], so we cannot do the race earlier, maybe a week, but that won't change a lot. Also there's the availability of the track. We do rent the track, as it is owned by the city and the park is a public park; it's open for the public during the summer, there's the beach inside the track, so we need to give the park back to the Montreal public.' That's also one reason why Montreal would be hesitant to shift to the end of the summer: the preparations would then take place in the summer months when the park has more visitors.

Because of its island location it is a compact venue, and the fast-flowing river is at points just a few metres behind the walls of the circuit. There are just two access roads on to the island and a convoluted one-way system for traffic, half of which is on a gravel track, and the first sign of congestion can easily trigger snafus. The car park is just a slim patch of hilly grass separated from the rowing basin by a car-wide strip of tarmac, which is used as the exit road. It has been known for someone to slip on the pedal, reverse too hastily and nudge a hapless passing motorist partially into the shallow basin.

It used to be a venue that had some of the weaker facilities, with just a single-storey pit building, creating cramped working conditions for mechanics. Hospitality units for each team were also cosy while the media centre bobbed up and down on a floating platform on the rowing basin. That was always forgiven, at least by the majority, because of Montreal being a great place for a race. Yet to secure a long-term contract Montreal needed to invest in the facilities and after the 2018 event the old paddock was torn down. Despite the tight timeline and the limitations of the harsh Canadian winter a swish three-storey pit building, complete with a wooden roof, was constructed just in time for the 2019 race – and guided Canada's grand prix into a new era. From the top floor there is a 360-degree view of the pit lane and paddock while across the Saint Lawrence the Montreal skyline provides a suitable backdrop.

'I think the location is very special. We are on an island, surrounded by water,' says Dumontier. 'It's probably, for us, the nicest image from the broadcast when you look at it, it's green, it's unique. We only have two entrances, only two bridges, to get on the island. The good thing is because the site was the Expo world fair in '67 they built the subway station, so 95–97 per cent of our clients are coming via subway. We don't have tunnels [on the island]. That island was man-created, that didn't exist before 1965, so when we built the new pit building we had to use pillars because when you dig more than two feet [60cm] you hit water – it's another challenge we face!'

The subway station, located on the adjacent Saint Helen's Island, takes visitors to the Circuit Gilles Villeneuve from the heart of Montreal. It is, bar the Old Town, not the prettiest of cities: parts are a little bit shabby and concrete-heavy, and the Metro is a haven of brutalist architecture, but given how half of the year is spent shielding from brutally cold weather it is little surprise that function is favoured over form. Because of that it has a labyrinth of passageways underground connecting several districts of the centre of the city to ensure that in the winter months its citizens can get around without ever having to venture into the freezing conditions.

Canada's event comes during a point in the year where the bulk of the narratives for the season have been cast. There have been enough sample sets to see who's competitive, who's in the mid pack, and who's at the tail end of the field. Yet whatever the order, and irrespective of the quality of the on-track product, Formula 1's news cycle keeps spinning.

As with every sport, Formula 1 has its broadcast media, and then also has its written media, who set themselves up in media centres, usually inside the main pit building (though rarely with an actual trackside view). Formula 1 controls the broadcasting licenses and access, but the print media and photographers come under the jurisdiction of the governing body, the FIA and its media team. There are approximately 300 permanently accredited journalists with a season pass, and for each individual grand prix there may be as many as 200 international requests and 50 or so national submissions. Numbers ebb and flow during the season, with European, American and opening/closing events tending to be the best attended, while those such as Australia, Azerbaijan and, increasingly, São Paulo tend to be quieter owing to travel costs, logistics and calendar placement. The FIA's responsibility also includes the bureaucratically tedious but necessary elements such as desk stickers, trackside tabards, pit lane access, car parking passes and grid access forms, as well as the need to deal with requests – and official statements – which will arise ad hoc.

'We try and create a consistent level of facilities, we define the media centre operations guidelines and we update it year on year depending on what the new requirements might be,' explains the FIA's Tom Wood, who spent five years as the Formula 1 Media Delegate. 'We try and ensure there's a standard that's upheld everywhere we go, so when journalists arrive they know they can work in an environment that has everything they need: catering, TV arrangements, internet. The local promoters have to create a media centre that fulfils the guidelines, as obviously there's more media coming to these events than any other local events. Then you need photographers, [and] we can split journalists and photographers as they can be on different schedules. We try and make it as comfortable for everyone. There's different infrastructure in place depending on where we go. Photographers once upon a time were shooting on film, a few photos a weekend. Now they send huge amounts of data, and everyone's streaming things, so there's a practical side which is defined and dealt with before we arrive.'

That happens early in race week, ensuring the majority of things are in place for Wednesday afternoon, ahead of matters properly kicking off with Thursday's media activities. Half of the 20 drivers will be present in an official press conference while the remaining half will have separate media sessions in team hospitality units. A team representatives' press conference, involving senior figures from different teams, takes place on Friday, while there are official conferences after qualifying and the grand prix for the top three finishers. The other drivers have to make themselves available in the mixed zone – located behind the TV crews – or in hospitality units, while increasingly teams also produce their own sugar-coated content.

'Pre-event I define the line-ups for the press conferences, liaising with the teams, to ensure everything is scheduled properly and they know where they're going and when,' Wood explains. 'We have regular meetings with team communication departments, and F1, to look at what we're doing and ensure we're doing the best we can for the teams and also that the media are getting as much access as is reasonable given that we have 24 races a year to manage. Even to the point of taking pictures of the route to the press conference room, as drivers don't know where they're going and can be late, as usually we run to quite a tight schedule. Our job as the regulator is to make sure everyone is where they need to be for those mandatory activities, as if someone is late it screws it for everyone else. We deal with the transcripts from that, as it's a permanent record of what happens and what's said.'

The FIA's media-related responsibilities also include chaperoning drivers immediately after qualifying and the race, managing the press conferences, and dealing with official pre- and post-race procedures, as well as issuing documents and notifications to teams and the media, such as drivers or teams being summoned for wrongdoing and the subsequent penalty. The FIA is also the port of call for official updates and information – usually when something is going wrong – during grand prix weekends.

'In F1 every weekend throws something up that's going to be different or will require reaction,' Wood says. 'So it's not just scheduling people and making sure they're there, it's: *OK, we've got an incident* or *We're launching this project,* so it's managing that around the normal day-to-day stuff. It's how full-on Formula 1 weekends are. It's why we do it. It's always exciting and it's just being there to manage it in the right way.'

The fast-paced nature of Formula 1 also makes it stand out from other industries.

'The reality check moment is when I meet other PR professionals, from banking, pharma, engineering, all big important things – we talk about these things, solving poverty, helping something medical – and [then] we're talking about cars going around in circles and you feel like, *OK, we love our sport and it's hugely important for us – but it is a sport,'* explains Will Ponissi, Head of Communications at Sauber. 'But for their campaigns and comms they're like, *We've been planning this for three months, six months, a year, how much time do you have to prepare the messaging?* And I'm like, *About 30 minutes between the end of the race, a story breaking, and being there with a reaction or a statement to a report* – so the challenge is maybe our topic is not the most difficult to talk about, but we have to do everything condensed, the rhythm is so much higher, the pace of everything we do is F1-quick, so that keeps it fresh and interesting.'

Canada is on Formula 1's calendar until at least 2031 after a long-term agreement was reached shortly before the pandemic. The Circuit Gilles Villeneuve will remain the venue despite some overtures to relocate to a different area of Montreal or even to a different city. It was a year's absence from the calendar in 2009, because of a commercial disagreement, which prompted a realisation among various parties, most notably those who were ambivalent towards Formula 1, and forced a rethink.

'First of all, the grand prix is the largest sporting event and tourism event in the country,' says Dumontier. 'In 2009, the Montreal people, Quebec, government, realised then what it means to have a grand prix. Back then, we were the only race in North America, so I said: "You cannot call that a world championship, there's no race in North America," so everybody in Montreal realised what it meant to have a race. There was a lot of pressure on the other side – teams, broadcasters – to go back to Montreal. It's a race appreciated by the F1 world.

'We're the oldest race outside of Europe in the championship. It's been in Canada for over 55 years now, at that track since '78. I see Montreal being on the calendar for a long time.'

Downtown Montreal is only a short metro ride from the Circuit Gilles Villeneuve

12

AUSTRIAN GRAND PRIX: RED BULL RING, SPIELBERG

'It's one of the shortest, if not the shortest, tracks we go to with just over one minute per lap, but it's a technical and challenging track, so it doesn't mean it's an easy circuit.'
NICO HÜLKENBERG

Formula 1 heads from the middle of Montreal to the verdant rolling hills of the Styrian Alps, a rural tranquillity that is shattered by the roar of a grand prix and where one organisation has given its country's motor-sport scene its wings.

There are some countries steeped in motor sport, with multiple charismatic front-running drivers, including those who have become World Champion. But there are only eight countries in the world to have produced more than one World Champion: the United Kingdom, Germany, Finland, Italy, Australia, Brazil, the United States and, the smallest of the bunch, Austria.

Jochen Rindt was crowned 1970 World Champion, the only driver to have been so posthumously, after a fatal accident at Italy's Monza. Niki Lauda accrued three world titles, in 1975, 1977 and 1984, during two spells of competing, survived one of the most brutal accidents in history at the Nürburgring in 1976, and became a key management figure of Mercedes' title-winning operation in the 2010s. Austria also had Gerhard Berger, who established himself as a front runner for over a decade through the mid-1980s to mid-1990s, winning 10 grands prix.

But the most influential Austrian, who had a transformative impact on Formula 1, was one who never went anywhere near a cockpit. Dietrich Mateschitz co-founded Red Bull, which makes the energy drink that has gone on to become a worldwide brand. It aimed its product at the young generation, using viral marketing, and swiftly moved into the sporting and entertainment arena through sponsorship and extreme branding exercises. Then it went even further.

Red Bull first became involved in Formula 1 in the mid-1990s as a partner of Sauber, and as a supporter of young drivers, which evolved into a fully fledged academy. When Ford pulled the plug on its underachieving Jaguar Racing team in 2004, Red Bull acquired the outfit. Red Bull arrived as a party team, shaking up the establishment, holding extravagant events, and trying to inject some fun into a sport that was regarded as straight-faced and sterile. But it had ambitions that were deadly serious. Within five years it had gatecrashed the usual front-running order and emerged as a long-lasting championship contender.

Within a year of taking over Jaguar, Red Bull had a second team, after buying the ailing backmarker Minardi and renaming it Toro Rosso as a nod to Minardi's Italian roots. Red Bull turned the midfield team into a title-winning operation within five years while Toro Rosso – which morphed into AlphaTauri and then Visa Cash App RB – was essentially the training ground for Red Bull's junior drivers, a public exam where they would either sink or swim. Succeed and promotion to Red Bull Racing was the prize – one that was scooped by the likes of Sebastian Vettel, Max Verstappen and Daniel Ricciardo. Fail and there was another junior motoring along the conveyer belt ready to step into the seat.

At roughly the same time as Mateschitz's empire expanded into ownership of two Formula 1 teams, his country's sole top-line motor-racing circuit was also available.

A fast and flowing circuit called the Österreichring, it was first opened in 1969 in the fertile Styrian hills, where somehow each valley conspires to be prettier than the last one.

The venue swiftly established itself on Formula 1's calendar. Rindt claimed the first ever pole, in 1970, but hopes of a home victory went up in smoke amid engine failure – and within weeks he was dead.

Austria eventually had a home winner in 1984, when Lauda triumphed during his third and final title-winning campaign, but the old Österreichring lasted only three more years before being deemed too old-fashioned for Formula 1's developing landscape.

Austria returned in the mid-1990s, when the country's telecommunications firm A1 invested, and the venue was imaginatively rebranded as the A1-Ring. Gone was the old sweeping Österreichring and in its place came a truncated version, bringing it up to modern safety standards, some of the old sections of circuit being bypassed and other sections modified. But the grand prix failed to make money and after 2003, at a time where Formula 1 was increasingly looking to fresh pastures eastwards, Austria fell from the schedule once more.

Mateschitz bought the A1-Ring in 2004 and had grand plans for the facility in Spielberg: the racetrack would be renovated and its facilities enhanced, with a racing school and hotel among the proposals submitted to local authorities. Yet shortly after the old facilities were torn down, and an enormous access trench carved through the track on the pit straight, the plans were blocked amid environmental concerns. 'Projekt Spielberg' looked dead in the water and Austria's once famous track resembled a wasteland, shorn of racing, but in 2008 permission was finally granted. A three-year construction period followed and in 2011 the Red Bull Ring, retaining the A1-Ring layout and utilising a similarly ingenious naming method, reopened. Formula 1 returned in 2014. It is a short circuit (just 4.3km/2.6 miles long), with the briefest lap time on the schedule (just a few seconds over one minute), and it has only 10 official corners – and only seven of these are recognised as an actual turn by engineers.

'There's a lot of nature and the circuit itself is built into the mountains,' says Nico Hülkenberg. 'There's great scenery, it's a scenic route in to the track, and it's very green. It's one of the shortest, if not the shortest, tracks we go to with just over one minute per lap, but it's a technical and challenging track, so it doesn't mean it's an easy circuit.'

It is indisputably Red Bull's facility and it is complemented by clinical Austrian styling and atmosphere. The grand Red Bull Wing, sculpted similarly to the rear wing of a racing car, perches over the pit straight. Within that facility Red Bull-branded merchandise is readily available within its shop, Red Bull-branded components are placed on show where possible, a plethora of historic Red Bull-branded machinery – such as old Formula 1 machinery and MotoGP-branded bikes – is dotted around the place (sometimes even corralled into narrow corridors), while fridges are regularly replenished with Red Bull and its startlingly vast range of products – try the apricot edition, the red edition (watermelon) and the ruby edition (spiced pear), and then see how bad the shakes are. The entire facility has been a boon for the local region: the circuit and its surrounding structures is a major employer, and Red Bull also heavily promotes Styrian (and wider Austrian) delicacies – ham, lamb, sweet treats such as apple strudel, and a plethora of pumpkin and pumpkin oil – and it is careful to source locally.

The Red Bull Wing also houses one of the best media centres of the year, located entirely across the top floor, with the floor-to-ceiling slanted glass-fronted structure providing views of the start/finish line and into the pit lane and team garages, as well as the majority of the circuit itself and the wider Murtal Valley.

Mateschitz's dedication to the area was such that a handful of nearby historic buildings – including castles, hotels and restaurants – were also acquired, some renovated and then associated with the circuit. Mateschitz, who shrank from public acclaim and kept his roots in his native Styria, died in 2022, but his empire and influence on Formula 1 remains enormous.

Red Bull's *no risk, no fun* ethos was instilled by Mateschitz, and his fingerprints are all over the team and the Red Bull Ring, meaning the continuation of the Austrian Grand Prix, through at least 2030, is something of a tribute.

'I think you feel his presence everywhere, whether it's the hotels you stay in, the investment he made,' Red Bull Racing Team Principal Christian Horner said at a press conference in 2023.

'I remember the delight he had at bringing this race back to Austria and the passion he just had for racing. Whilst not here in person, his presence, you can feel it everywhere.'

Red Bull's landmark at the Red Bull Ring is the leaping bull. A 12m (39ft)-high steel sculpture, of rust colour with gold horns, the bull is moulded leaping through an aluminium arch, located on a raised grass mound in the centre of the track. But in typical Formula 1 fashion, one brand being so prominent leaves itself open to friendly teasing.

Some rival teams, notably Mercedes, will never refer to the Red Bull Ring in promotional material, instead calling the venue 'Spielberg'. And when Mercedes comfortably claimed a 1-2 finish in the Red Bull Ring's debut race in 2014, a few giddy mechanics headed over to the bull sculpture and cheekily attached Mercedes' three-pointed star logo. The advert released in newspapers in the aftermath? *Silver Arrows: gives you wings*. By the time Formula 1 rocked up in 2015, a temporary security fence had been installed around the bull to prevent a repeat. Mercedes triumphed for four successive years before Red Bull finally scored victory at its own facility in 2018.

The Red Bull Ring's rural setting, and the circuit's facilities, played an influential role in the early stages of the pandemic. Formula 1 reached an agreement to host the opening two rounds of the delayed 2020 season in Austria – in early July – at the Red Bull Ring, marking the first time in history that one venue held multiple races in the same championship. One retained the national moniker while the second became the Styrian Grand Prix, after the local region. The FIA compiled a detailed Covid-19 Code of Conduct document which each person had to sign. The bucolic venue was closed to fans and only essential team personnel were present, each subject to stringent and regular Covid tests. Strict biospheres were enforced, the team personnel not permitted to interact with other teams or even different departments of their own team, and distancing was implemented where possible. The nearby Zeltweg Airfield – the home of the Austrian Air Force, who sometimes perform ear-splitting flights early in the race weekend – transported personnel in and out, meaning there was little risk of interacting with large population areas, reducing the risk for both Formula 1 and the local community.

There were no grandiose motorhomes, with drivers instead in Portakabins in the paddock, while the usual VIP areas were repurposed for team personnel. There were no fans, no Paddock Club, no sponsors on the ground – and consequently no marketing personnel – while the Red Bull Ring utilised its own facilities to provide catering for teams in that repurposed space. TV crews were allowed in limited numbers and interview areas expanded, while all other media activities were conducted virtually. The limited number of present journalists were confined to the media centre. Each team had a couple of photographers, mandated to wear team uniform, while other snappers were permitted to shoot only trackside. Engineering meetings, driver briefings and other in-person gatherings were carried out online, while other aspects of the weekend were either dropped, such as the drivers' parade, or modified, most notably the podium ceremony. Enormous sponsorship banners covered some of the empty grandstands, while selective camera angles partially hid the absence of spectators. The sound of the V6 power units also meant Formula 1 suffered less than other sports, such as football or tennis, which took place in echoey, empty stadia. Off-track it may have resembled an insipid test day, but the championship was able to begin and the made-for-TV product was essential during a period of previously incomprehensible uncertainty.

This was a strong marketing boon for Red Bull – being able to run two events in a safe environment – while Austria also profited by signalling it was open for business. Formula 1 was able to use the events as a template, showing other governments and sports around the world that it could act responsibly, shield the local population and not be a drain on resources – a crucial takeaway for a sport heavily reliant on international travel. The wheels of the championship turned once more and essential contracts were able to be fulfilled with the eventual completion of a 17-round season, at 14 circuits in 12 different countries, in the space of six months.

Fortunately fans were able to return for 2021's pair of rounds in Austria, the circuit once more filling the breach when travel restrictions precluded grands prix in Australia, Canada, Singapore and Japan – and its green scenery modified by an influx of orange. The rise of Max Verstappen, and his intrinsic association with Red Bull, has made Austria's GP round one of the home events for his travelling fans. Verstappen's Orange Army turn the Austrian GP into a vibrant event, letting off smoke flares at opportune moments, and playing loud club music in the busy campsites at all hours of the day. Many make the lengthy road trip to and from the Netherlands, frequently in caravans and campervans, some of which are elaborately decorated to show their devotion to Verstappen. And after some wayward stumbles back to the campsites on Sunday night, it's a long and slow voyage back to the Low Countries on Monday.

The rustic setting of the Red Bull Ring means that there is little in the way of international hotel chains. The closest major conurbation is Graz, Austria's second city, known for its historic centre and student lifestyle, but that is still an hour away from Spielberg. Vienna, Austria's beautiful capital with its charming cafés and classical music halls, is two hours' drive away, at least. Many personnel opt to shirk hotels in favour of local guesthouses, holiday homes typically used for skiing getaways in the winter, or even residential facilities on farms that are dotted around the region. That gives the race weekend a distinctly traditional and old-school flavour – some liken it to a detox, since attendees find themselves ambling about creaky wooden houses, grappling with patchy Wi-Fi and using quaint equipment rather than swish state-of-the-art hotel facilities. Those based near farms can be awoken by the sound of clanging cow bells, the midsummer dawn chorus arriving outrageously early, or tractors working at similarly early hours across the fields that stretch out for miles across the countryside. There are some communes where it feels as if nothing has changed for a hundred years. On sunny and hot days, the Styrian hills offer total tranquillity, though on rainier and cloudier days the scenery is shrouded in mist and the notion of such bucolic escapism is a little less endearing. The drive towards the circuit passes by fields and verges festooned with buttercups and poppies and past chocolate box houses. It feels as far removed from some of Formula 1's newest venues, such as Saudi Arabia and Qatar, as is humanly possible.

The success of Austria's return, facilitated by Red Bull and entwined with the rise of Verstappen, highlights the demise of its linguistic ally and neighbour, Germany, a country renowned for its automotive industry. That nation was once a staple of Formula

1's calendar: lengthy blasts between the forests of Hockenheim, and the epic voyage through the 100+ turns of the narrow and fearsome Nürburgring's Nordschleife (dubbed the Green Hell), made Germany a firm fixture. In fact, it has hosted 79 grands prix, a figure that – as of 2024 – has been surpassed only by Italy.

The sport blossomed through the mid 1990s amid the emergence and popularity of working-class hero Michael Schumacher and the country's reunification. Hockenheim, located near the picturesque town of Heidelberg, was known for its high-speed blasts on a track carved between the forests, though its layout was truncated and modernised in 2002. Hockenheim and the Nürburgring – in its GP Strecke form rather than the unfit-for-purpose Nordschleife – both held races annually for a decade before entering a rotation arrangement after the eventual seven-time champion Schumacher bade farewell. But amid dwindling attendances, a lack of promotion and funding deficits as the money needed to hold Formula 1 events rose, Germany's round twice dropped from the schedule – in 2015 and 2017 – and then Hockenheim's contract finished in 2019. The Nürburgring was revived in 2020, as a one-off event in the pandemic-ravaged calendar, and did so under the regional Eifel Grand Prix moniker. But that was Formula 1's last foray to a country synonymous with both the automotive and motor-sport world.

'The main reason is there is no funding for it,' says Christian Menath, Formula 1 journalist for Germany's *Motorsport-Magazin*. 'Hockenheim or Nürburgring, wherever, they'd have to fund the race completely themselves – no public money – they are trying to get public money, but it's not working in Germany, as it's not in vogue. We have a Green Party involved in the government, so it's very unlikely to get any money for a Formula 1 race, but it was the case in the old government as well – it's just not a project where you can spend public money at the moment in Germany.'

Since Schumacher's success Germany has had champions in the form of Sebastian Vettel and Nico Rosberg, and statistically the country is the second most successful in terms of titles and victories. But while both Vettel and Rosberg had a hardcore fan base, neither grabbed the country in the same way as Schumacher.

Both have since retired and there is no German megastar rising through the junior categories.

'In 2023, in Germany, it was all pay TV, and Germans are not willing to pay for TV, so the figures were the same as Austrian TV – where it's free to air – and Austria is a tenth of the German population,' says Menath. 'There isn't interest at the moment and it's a pity. But even during the Vettel/Ferrari years . . . you had a German driver at Ferrari, fighting for the championship, and the race was still not sold out – that says quite a lot. If you look at the grandstands the last few races we had at Hockenheim, they were covered with Rolex banners because they didn't sell enough tickets, which is quite embarrassing.'

It is a shame that a country with such rich history in Formula 1 – and which felt like a cornerstone of the championship during the Schumacher years – has suffered such a decline. It also acts as a warning to other historic venues: nothing lasts forever and events – and fans – must never be taken for granted.

Austria is the most rural event on Formula 1's modern calendar

13

BRITISH GRAND PRIX: SILVERSTONE

'The track is fantastic – it has some of the greatest corners on the calendar, which I really like and it's super high-speed.'
                                                            NICO HÜLKENBERG

'The energy here is just spectacular and there's nothing quite like it anywhere else,' says Lewis Hamilton.

The expansive Silverstone venue, roughly halfway between London and Birmingham, is where it all started for the Formula 1 world championship, on 13 May, 1950.

Silverstone is the home grand prix of seven of the ten teams – and by extension where the bulk of personnel stay – and of Formula 1 itself, which is based in Britain and has its main offices in St James's Market, just a short walk from Piccadilly Circus in central London.

Britain has a rich history in Formula 1. The country has produced more drivers than any other nation in Formula 1 history: 145. More race wins: over 300, which is 100+ more than any other country. And, unsurprisingly, the most titles: 20, split between 10 different drivers. From the days of Mike Hawthorn via Graham Hill, Jim Clark, John Surtees, Sir Jackie Stewart, James Hunt, Nigel Mansell, Damon Hill, Jenson Button and Lewis Hamilton, there has always been a flow of talent in Britain. The presence of race winners Lando Norris and George Russell, and the emergence of Oliver Bearman, bodes well for Britain's Formula 1 future. Not since the early 1950s has there been a Formula 1 round without a British entrant – discounting the period when the Indianapolis 500 counted for the world championship but Formula 1 regulars did not enter.

Britain is one of only two countries – along with Italy – to feature in each world championship season, with the grand prix held permanently at Silverstone since 1987. A relatively simple and flat course laid out inside Aintree Racecourse, in Merseyside, hosted five races in the late 1950s and early 1960s, while Kent's high-speed undulating Brands Hatch alternated biennially with Silverstone from 1964 to 1986. Donington Park held a one-off and rain-hit race in 1993, famous for Ayrton Senna's stunning first lap and victory, under the European GP title. Senna rose from fifth on the grid to sweep past esteemed rivals, including Alain Prost, Damon Hill and Michael Schumacher, and into the lead before the end of the opening lap, going on to win the race by 83 seconds.

Donington Park attempted to wrestle the British Grand Prix away from Silverstone for 2010, signing a long-term contract with the promise of reprofiling its track, but the venture was a bust and Silverstone duly retained host rights.

The Silverstone circuit was formed in the aftermath of the Second World War, on the site that had been RAF Silverstone. The base, formerly agricultural land, was created in 1943, and featured three runways as the location for the Operational Training Unit of No. 17 Group.

RAF Silverstone was decommissioned after the war and in 1947 a few local motor-sport enthusiasts recognised the potential for a track. They worked their way into the locked venue and 12 cars raced around the airfield; one of them clipped a sheep and thus the event unofficially became known as the Mutton Grand Prix.

In early 1948 the Royal Automobile Club was searching for a venue at which to revive grand prix racing in Britain after the war. Donington Park, the only permanent pre-war venue, had been requisitioned during the war and was not in a state to hold motor racing, while Brooklands, in Surrey, had been largely built over. There were several abandoned airfields and a couple were considered before Silverstone was selected. That October a grand prix was held, heavily utilising two of the runways and some of the perimeter roads that encircled the airfield, and more than 100,000 spectators were in attendance. For 1949 the layout was modified, the runway sections being removed and the perimeter roads broadly making up the new Silverstone circuit. It is far from the only circuit in the UK to have its foundation in the war: Snetterton, Thruxton and Croft are among venues to be borne out of the runways and access roads of airfields.

By the end of the late 1940s motor racing had been firmly re-established following its interruption by the war. Motor racing may have begun in the 1890s, with the first 'grand prix' taking place in 1906, and there had been an array of short-term championships, but only in 1950 was a structured world championship created – run to Formula 1 regulations introduced in 1946 – and on 13 May the opening round of a seven-event season took place at Silverstone. Among the attendees were King George VI and his wife, Queen Elizabeth (later the Queen Mother), who witnessed Alfa Romeo's Nino Farina lead home a 1-2-3 for the fabled Italian manufacturer.

Silverstone's layout has evolved over the decades, one of the biggest changes coming in 2010/11 when a new section of circuit was introduced, and the start/finish line and paddock were relocated. That coincided with the construction of The Wing, a vast building that stretches the length of the Hamilton Straight, and which was a much-needed upgrade to the fading facilities in the old paddock.

Yet the fundamental characteristics of Silverstone remain unchanged: swathes of its layout can be traced to its 1950 iteration and it is still a high-speed challenge. Most of the original runways are still intact and easily identifiable from aerial photographs, and are used as access roads, parking areas and – since the 2010 revamp – even the track itself. The renamed Wellington Straight, named after the Second World War aircraft, was introduced as part of that modification.

'The track is fantastic – it has some of the greatest corners on the calendar, which I really like, and it's super high-speed,' says Nico Hülkenberg.

Silverstone has a high-speed flow, with some corners such as Abbey and Copse taken in eighth gear at almost 300km/h

(186mph), while the Maggotts/Becketts/Chapel corners combine to create a ribbon of tarmac that features staggering changes of direction, as drivers sweep left, right, left, right, with barely a dab of the brakes and a flick down of the gears.

It is one of the easiest yet complicated weekends of the year for the bulk of the teams. Silverstone's location means forsaking the usual extended travel arrangements, but while some personnel enjoy staying at home, others struggle outside the usual travel structure – and the associated external distractions of a normal life – during a grand prix weekend.

Silverstone sits in the heart of Britain's motor-sport valley, an area which broadly stretches along the M4/M40 corridor to the north and west of London. Its presence and expansion have been a self-fulfilling prophecy. Formula 1 grew through its early decades, and consequently so did team sizes, and as a result so did the factories required to house personnel and the increasing range and levels of machinery needed to compete. Most teams began life as a smattering of enthusiasts keen to go racing, some on a wing and a prayer, surviving hand to mouth and driven purely by the passion and willpower to succeed. And while that hunger to triumph remains, the 10 teams on the grid are now all worth over $1 billion, even the smallest employ hundreds of people, and they are the cornerstones for promoting products – be it energy drinks or supercars – on a global scale.

Each Formula 1 team has its own headquarters where the cars are designed, constructed, maintained, developed and fettled. These facilities have state-of-the-art equipment, such as wind tunnels, simulators and remote operations providing assistance to those at the circuit – and this is where the majority of personnel work throughout the year.

Red Bull may be an Austrian corporation, but Red Bull Racing is based on an industrial estate in Milton Keynes, on a campus that has expanded to include Red Bull Advanced Technologies, its engine division Red Bull Powertrains, and its MK7 conference centre which showcases its Formula 1 machinery. Similarly, Mercedes, the German manufacturer, has its Formula 1 headquarters a few miles from Silverstone, on an expanding leafy campus in Brackley, where its Applied Sciences division is also installed. Mercedes' engine facility, Mercedes AMG High Performance Powertrains, is located just outside Northampton, in the village of Brixworth.

McLaren's futuristic circular Technology Centre, which emerges from the ground like a spaceship perched next to a lake, is beneath the Heathrow flightpath in Woking, while Williams' factory is in the village of Grove, in the Oxfordshire countryside, having moved across from nearby Didcot in the mid 1990s. Williams also has its Heritage Museum on its base, where an array of its iconic, title-winning machinery is on display; its most recent title-winning car, from 1997, hangs from the lobby in reception.

Alpine, through its many iterations – Benetton, Renault, Lotus and then Renault again (and by the time of publication quite possibly another name) – has long been based in twee, rural surroundings in Enstone, where the officially titled Whiteways Technical Centre is located on the site of a former quarry. Renault's engine division, meanwhile, is based across the English Channel at Viry-Châtillon, near Paris. Haas competes under an American license, but its race operations and core functions are in Banbury, while RB, which has its main factory in Italy's Faenza, has an aerodynamics facility in Milton Keynes, having relocated from Bicester to be closer to parent company Red Bull. Aston Martin, stretching back to its infancy as Jordan Grand Prix in the early 1990s, has been based at Silverstone, directly across the Dadford Road roundabout from the circuit's main gate. When it deemed, rightly, that it had outgrown its old factory, an entirely new facility, called the Technology Campus, was constructed on acquired land next door and opened in 2023.

'It's right across the road,' says Aston Martin driver Lance Stroll. 'So it doesn't get much more of a home race for the team than this one! It's great. I mean, just for everyone in the team, there's a lot of guys over at the factory that don't really get a chance to come, they get to watch some racing, have a chance to come over and see the car and get a feel for the race weekend, which is cool.'

Beyond the Formula 1 teams, a plethora of junior teams, including Silverstone-based Hitech, are located in the area, along with suppliers and contractors. This is a world-leading region for motor sport, engineering and associated knowhow, and an estimated 40,000 people are employed within the chain. It is also a region that drivers know well from junior days, several having relocated to the UK as teenagers – including Oscar Piastri and Zhou Guanyu – and others brought there by the team. Red Bull housed members of its junior team close to its factory, in Milton Keynes, which was a culture shock for some.

'I'm not saying I'm not a fan of the UK,' emphasises Yuki Tsunoda. 'I'm not a fan of Milton Keynes!'

'I actually grew up around this area,' says Alex Albon. 'I learned my ARKS test, which is how you get your racing licence, I did it here. So you could say this was the very start of my racing career. I learned how to drive. They asked you: *What does your right foot do? What does your left foot do?* All that stuff!'

For the non-Brits of the paddock, Silverstone can be a bit less enticing, given that the budget hotels of Northamptonshire are not the most exotic part of life on the road, but this is still an iconic grand prix.

Silverstone is the closest that a Formula 1 event gets to a music festival. The fields surrounding the racetrack are transformed into campsites for the weekend, and the local villages are swamped by enthusiastic spectators. They load up with fold-up chairs, flasks and umbrellas, with sun cream and pacamacs, braced for any and every weather scenario that the great British summer can deliver, and queue outside the gates to the sound of the dawn chorus to grab the prime real estate on the grass banks. Silverstone's surroundings resemble Glastonbury, with a plethora of tents and flags dominating the skyline, and the smell of meat on barbecues wafting through the air. Some years a summer heatwave means the grass has long turned a yellowy brown, while in other years the fields have already become a quagmire by the time the weekend's action begins. There is usually broad support for teams and drivers up and down the field, but historically there is widespread backing for the home representatives. Nigel Mansell won four times, Jim Clark five, and Lewis Hamilton has hoisted aloft the glistening golden Royal Automobile Club trophy on nine separate occasions, a record for one driver at one circuit. Hamilton won the British Grand Prix for the first time in 2008, mastering the wet conditions to such a degree that he was victorious by 68 seconds, and won the race in 2014, 2015, 2016, 2017, 2019, 2020, 2021 and 2024 – the ninth also ending a two-and-a-half-year win drought for Hamilton.

'This grand prix is the best,' said Hamilton in a press conference in 2023. 'I remember coming here the first time when I was a lot younger, racing in cadets. I think it's around like the Brooklands area, they had, like, straw bales. I remember crashing! It wasn't a good race. But then, racing, coming here for my first time in Formula Renault. And coming and watching the race, I think for the first time – I must have been like 13 or 14 or something – with McLaren, and just standing at the back of the garage, dreaming of one day being in the car, and then getting into Formula 1 and having the first year and it's been [a] phenomenal journey that I've had here. I remember the crowd the first time, in 2007, when I got the pole position. I think that's probably the only time I've ever heard the crowd over the sound of the car. So that's pretty special.'

Silverstone's attendance has now reached almost half a million across the course of the weekend. Approximately 160,000 people present around the vast facility on race day alone – the highest race day attendance of any race, and a remarkable figure considering the location and the high price of tickets. Even the very cheapest race day tickets are well into three figures, which is a shame because it can dissuade some from attending, but demand still outweighs supply.

'I think, overall, the crowd just lifts you and the whole team up,' said Hamilton. 'And there's this kind of feeling of floating when you come here on that positive energy that everyone brings, and just seeing so many flags, so many caps and people supporting us. It really just helps spur you along. And when you're in the car as well, when you see them, you can see them through so many of the corners on the outside of the circuit and it's really just really encouraging.'

Lando Norris became the youngest Briton to race at the British Grand Prix in 2019, aged 19, and he has since gone on to take multiple podium finishes at Silverstone.

'It's very intense,' he says of the British Grand Prix. 'I've had to calm some of it down because like the first few years was just too much. And that's almost when you need it to be the least! And maybe now I can deal with more than what I can do or could have done. It's a lot, just because there's a lot of family, friends, sponsors, partners, and it's the one race a year where you also don't mind going above and beyond just to do more stuff for the fans. You go and do meet-and-greets later in the evening and things like that. You still have to control it because you don't want it to just be chaos and then take anything away from the fact of why you're there in the first place. But yeah, it's the one race where I'm like: *You know, if there's something in the evening . . .* and I'm like, *Yeah, I want to go and do it*, just because it puts a bigger smile on your face than not. Like, no matter, anything you do there pretty much puts a smile on your face. It's always good fun.'

Silverstone has enhanced its off-track product in recent years, with music acts such as Kings of Leon and Stormzy having played the main stage in an area repurposed from the old Bridge turn, a fearsome curve that was bypassed when the revised section of track was introduced in 2010. The main stage is also the place for interviews with key figures, including drivers, who hang around after the race on Sunday to greet the thousands of fans who stay on.

Silverstone's layout places an emphasis on aerodynamic prowess: the majority of corners are taken at high speed, including famous curves such as Abbey, Copse and the Maggotts/Becketts complex, where the change of direction on a qualifying lap is staggering. As a former airfield, Silverstone is open and exposed, meaning it is susceptible to gusty conditions, even on ostensibly calm days. The wind can often change direction swiftly, which adds another challenge for drivers because this can alter the car's balance and behaviour, particularly a crosswind, and drivers must react accordingly. The high-speed corners, and the time drivers spend in them, makes Silverstone a high-energy circuit for the tyres, meaning preservation is a key element of the race. That was most prominent in an astonishing finale in 2020, when several drivers suffered late punctures, including leader Hamilton on the last of

52 laps. Hamilton completed half of the final lap on three working tyres as his left-front gradually deteriorated, but he somehow held on to claim a memorable victory. The only shame was that, due to the pandemic, there was no one in the crowd to witness such a feat. Fortunately, one year later, Silverstone was packed to the rafters for one of the all-time iconic Formula 1 moments, when title rivals Hamilton and Max Verstappen clashed on an extraordinary opening lap.

'It was the first race that was fully back vivid at post-pandemic capacity, so it was like F1 restored to its grandness, and there was just a great feeling of euphoria, good times were back again, amidst this amazing title battle between Verstappen and Hamilton,' says Formula 1 commentator Alex Jacques. 'I thought about where they were going to hit each other if they did hit each other on the first lap – one was Brooklands and one of them was where they did at Copse! To get a lap of such aggression, to see an illustrated heavyweight title fight, and the visual language of the cars and the driving to match that, it feels so intense, the risk – always there in F1, but it was apparent there from the very first corner.'

Silverstone's placement during the height of the summer means it is usually one part of a packed British sporting calendar. British Grand Prix weekends have previously been spent hooking up TVs or tablets to track progress during the FIFA World Cup or UEFA European Championships. Wimbledon often overlaps with the grand prix weekend and epic tennis matches have been played out while the cars are hurtling around Silverstone. Occasionally the Ashes are scheduled for the same period, prompting intense distractions for the British and Australian contingent and extreme confusion for the rest of the world: *What does 204-7 mean? What do you mean by* leg slip? *And, my word, one match – sorry, a* test – *lasts five days and there's still a chance no one wins?*

Occasionally the concept of a London grand prix in the city is mooted, usually during the build-up, ostensibly to raise awareness that the British Grand Prix is coming up. Proposals, usually bordering on the fantasy, crop up, typically incorporating some of London's most famous landmarks. Formula 1 took over part of central London in 2004, and again in 2017, for a demonstration run prior to the British Grand Prix. But an actual grand prix in the capital is unrealistic, given the disruption and expense that would be involved in bringing the championship to London, and the infrastructure required. It is nice and fluffy PR, but it is something London neither needs nor wants. Besides, 130km (80 miles) up the M40 is one of the finest circuits in the world.

In 2024 a new 10-year agreement was reached, securing Silverstone's place on Formula 1's schedule until at least 2034. That allayed fears in some quarters that Silverstone risked dropping from the schedule – nothing is ever assured until the ink is dry on the contract – and safeguarded a circuit that is one of the best in the world, adored by drivers and spectators alike. That long-term stability is allowing Silverstone's officials to continue investing in the facility. A new kart track 1400m (4600ft) long, capable of holding international-level karting events, is set to be constructed on the old Bridge corner, and further plans are afoot to help turn Silverstone into a year-round tourism and events venue.

In announcing the deal, Formula 1 CEO Stefano Domenicali described Silverstone as 'an iconic venue at the heart of F1 history, and as it approaches its ninth decade hosting grands prix, the event continues to attract fans from around the world for fantastic racing on track and the amazing fan experience off it.'

That is undeniable. Silverstone will always be the track at which the journey began for the Formula 1 world championship – Round 1 of over 1100 and counting. It would feel wrong for Britain, and Silverstone, to ever be overlooked on the schedule.

14

BELGIAN GRAND PRIX: CIRCUIT DE SPA-FRANCORCHAMPS, STAVELOT

'It's a pure racing track. It's old school, no room for error, and it's still very much a driver's circuit.'

STOFFEL VANDOORNE

Buried in the heart of Belgium's tranquil Ardennes is a ribbon of tarmac that rises and falls 100m (330ft) across its flowing 7km (4.3 miles), traversed by all the legends of the sport, and universally adored: the Circuit de Spa-Francorchamps.

After the conclusion of the First World War, the proprietor of the Liège-based newspaper *La Meuse* wanted to revive La Meuse Cup, a race which had been halted by the conflict. Jules De Thier met with racing driver Henri Langlois van Ophem and Baron Joseph de Crawhez, the mayor of the nearby thermal town Spa, to discuss the notion. They looked at the undulating public routes through the Ardennes which connected the villages of Francorchamps, Malmedy – transferred from Germany to Belgium after the war – and Stavelot. Those roads formed a near-triangular shape and it was deemed an ideal location for racing.

In 1921 racing got underway at the Circuit de Spa-Francorchamps, a trek of 14km (8.6 miles) around the Ardennes Forest – but for motorcycles, because only one car registered for the prospective race. A year later the Royal Automobile Club of Belgium organised the first car race at Spa-Francorchamps and in 1925 the Belgian Grand Prix was born, won by Alfa Romeo driver Antonio Ascari – father of future World Champion Alberto.

Racing was interrupted during the Second World War, when the Ardennes Forest was the focal point of the Battle of the Bulge. It is hard to envisage such luscious and peaceful forests being the scene of bloody and brutal massacres, yet racing returned to Spa-Francorchamps in 1947 and three years later it was part of the inaugural world championship season in 1950. Even in the early years of Formula 1, when many courses utilised public roads with little in the way of safety features, Spa-Francorchamps had a fearsome reputation. It was very fast, very narrow and very dangerous. If a driver went off, they were met not with a gravel trap or sophisticated barriers but with barns, barbed wires and even sheer drops down grass banks. In 1960 Chris Bristow and Alan Stacey were killed in separate accidents during the Belgian Grand Prix, while Sir Stirling Moss and Mike Taylor were injured in practice accidents.

The landscape changed in 1966 following a serious accident. Sir Jackie Stewart crashed in wet conditions at the Masta Kink and was trapped for almost half an hour in his upside-down car – which was leaking fuel. Stewart had to be rescued by contemporaries Graham Hill and Bob Bondurant owing to the absence of trackside crews, and was then mishandled by several medical crews, including by one ambulance driver who got lost. The debacle would have been slapstick were it not for the gravity of the situation and this sparked Stewart's crusade to improve safety conditions in Formula 1 across the following years. That included the introduction of full-face helmets, seatbelts, proper medical facilities and training at tracks, and the enhancement of safety structures at circuits.

The old Circuit de Spa-Francorchamps was deemed unsuitable and the race was cancelled in 1969, when changes that had been requested were not made. The following year it dropped off the schedule when temporary fixes were still deemed unsuitable. Spa-Francorchamps remained in existence for other categories, most notably sportscars, while Formula 1 briefly relocated its Belgian Grand Prix to a flat and featureless circuit in Nivelles, before establishing a home through the 1970s at Zolder.

Spa-Francorchamps' original track was finally considered unfit for purpose and revisions were carried out for 1979. Sections of the old layout, nearer Francorchamps, were retained and tweaked while the lengthy sweeping curves towards Malmedy and Stavelot were bypassed by the construction of an entirely new section of track.

The abridged circuit was reintroduced in Formula 1 in 1983 and in 1985 – after the race was initially postponed due to the new track surface breaking up – it was re-established as the home of the Belgian Grand Prix. Sections of the track remained public roads until as recently as the early 2000s, when a new road bypass was built and the facility was permanently closed off. Swathes of the pre-1970s circuit, including the sections towards Malmedy and Stavelot, remain public roads and are still driveable.

It remains a drivers' favourite. At 7km (4.3 miles), the layout is the longest circuit on the schedule and sweeps through the Ardennes, with high-speed corners, rapid changes of direction and greater elevation than any other venue on the schedule.

'Of course, I'm Belgian, so it's a home race, but I see it as a very authentic Formula 1 circuit,' says local Stoffel Vandoorne, who raced for McLaren in 2017/18. 'It's a pure racing track. It's old school, no room for error, and it's still very much a driver's circuit and everyone enjoys racing around there. I'm sure if you ask most of the drivers on the grid they rank it pretty highly. It's one of those that's always in the top three of the driver circuits. Maybe it's not the most glamorous race, but it has got its own character: it's in the forest, it has its own elevation changes, the fans there are pure racing fans.'

The first and final sectors are almost entirely flat out, with the opening parts of the lap featuring the Eau Rouge/Raidillon complex, one of the most renowned corners in motor sport. It is a section of track along which drivers gradually descend before hitting a compression and ascending a steep curve flat out – akin to scaling a tarmac waterfall – and is named after a nearby stream that has a reddish tinge due to iron-oxide deposits. The track rises 40m (130ft) at a peak gradient of 17 per cent, taken in under a couple of seconds in Formula 1 machinery, then the drivers traverse the crest and continue on the lap. The sight of 20 Formula 1 cars snaking through the complex remains majestic.

'Spa is a track I know well and I made my F1 debut here,' says Esteban Ocon. 'Spa is an iconic circuit and many drivers really love coming here, including myself. The most famous part of the track is probably the run from La Source down to Eau Rouge, up through Raidillon and along the Kemmel Straight towards Les Combes, which offers some exciting, full-throttle racing.'

The Eau Rouge/Raidillon complex nonetheless invokes mixed emotions and remains a contentious talking point in the modern era. The progression of Formula 1 cars and evolution of the corner mean that, while it is a spectacular sight, drivers easily take the corner flat out. It has been the scene of several major accidents and is regarded as one of the most dangerous pieces of racetrack anywhere on the calendar. In 2019 Formula 2 driver Anthoine Hubert was killed at the complex. He crashed to avoid a slow-moving rival and his damaged car was then struck at high speed by Juan Manuel Correa, who sustained life-changing injuries in the impact. It was the first fatality of a driver at a race in 24 years.

Hubert was a close friend of Pierre Gasly, who lays flowers at the scene of the accident every year.

'I think I relate [to] a place with emotions,' said Gasly in 2023. 'I've had the worst emotions of my life here [at Spa-Francorchamps]. And at the same time, it's one of my favourite tracks. It's very contradictory. I love this track, I love racing at this track, but at the same time I'll never forget what I felt going down these stairs [in the paddock] when my parents told me the news. It's obviously tough. But I accept the sport that we do and it's things you've got to live with.'

Another of Hubert's friends, Charles Leclerc, went on to claim his maiden F1 victory the day after the accident.

'It's always special to come back to the place where you've won your first race,' said Leclerc. 'It is a very special memory for me. It was a very special moment at the time, even though it was done in circumstances when it was obviously difficult to enjoy the moment, as we had lost Anthoine the day before.'

The tragic accident which befell Hubert was a confluence of highly unfortunate events, and everything came together at the same split second to produce a devastating outcome. There is no romanticism to the loss of a young life in pursuit of a dream: any death at a racetrack is violent, soul-destroying and brutal.

It was a reminder that the quest to improve safety can never relent.

The FIA has a Safety Department that is constantly undertaking research and development into new and evolving technology, while after every major accident there is an investigation to understand if anything can be learned and applied for the coming years.

There have been major developments, such as seatbelts, the Head and Neck Support Device (HANS) and, more recently, the halo, the cockpit protection structure that became mandatory in 2018. Some developments can be met with resistance, owing to aesthetics or comfort, but very quickly new practices become accepted and normalised. There have also been more subtle but important safety devices in recent years, such as an in-ear accelerometer – a sensor that detects motion; biometric gloves – which monitor vital signs via a sensor; and the duration for which equipment must be flame-resistant. The halo also features a high-definition camera pointing at the driver, which acts as a sort of black box in the event of a serious accident.

Drivers wear one-piece overalls, extending from the neck to the ankles, and have shoulder straps to assist with extrication. The biggest consideration is weight: drivers need them to be light and breathable, but there are tight regulations concerning fire resistance. There are also cosmetics in mind given that overalls double up as a walking billboard and there are sponsor and partner commitments. Drivers also have specially designed shoes, which can be as light as 200g (7oz). Helmet manufacturers must also meet stringent standards – crush protection, ballistics, fire tests – while drivers will have around three helmets per weekend available. Each can cost around $15,000 and they are the one aspect that drivers get to personalise, with their own design becoming their effective representation. Some drivers will create one-off designs for special events, such as their home grand prix. Even so, this must be done with lightweight paint that complies with the FIA safety standards.

The FIA has its own medical team, who work at each grand prix with local doctors and crew. There are also the medical cars, either a Mercedes AMG GT 63 S or an Aston Martin DBX707, and there are two on location each weekend. These high-performance vehicles are large enough to contain the necessary medical equipment, rescue devices and communication tools to follow the broadcast. The medical car trails the Formula 1 field on the opening lap, meaning they are quicker to the scene of any accident, before peeling into the pit lane, where the car remains primed to leap into action if required. There is an hour of track time set aside on Thursday afternoon for the safety car and medical car drivers to acclimatise themselves to the circuit, while also clocking several shortcuts and quicker access roads that may be used in an emergency.

Each circuit has a medical centre, which must maintain FIA standards, while the governing body has designated local hospitals. There is also a medical helicopter and if that cannot fly due to weather conditions – usually fog – then the local hospital must be reachable by road within 20 minutes. If neither of those boxes are ticked, track activity cannot take place. The extrication process – removing a driver from a wreckage in the event of a major accident – is meticulously outlined and practice takes place on a Thursday, with particular focus placed on the spine and neck position.

There are also stringent and detailed regulations concerning elements such as accident barriers and materials, debris fences and trackside lighting panels.

The work is constant, and often goes on under the radar, but there are numerous cases of safety advancements preventing or lessening the extent of injuries, or worse, while lessons can also be applied to the automotive industry. It is a grimly accepted, though oft unspoken reality that one day there will be a freak incident, or an unavoidable set of circumstances, which results once more in a death. But fortunately the work of the governing body, countless organisations and the push from inspired individuals across the decades – most notably the likes of Sir Jackie Stewart and Professor Sid Watkins – means that drivers have been able to walk away from situations that would have been previously unsurvivable.

Accidents are also more likely to occur in low-grip conditions and at some events particular attention is paid to the weather radars.

Spa-Francorchamps' location deep within the Ardennes means it is susceptible to chilly and wet weather even at the height of the summer. Entire days can be spent mired in drizzle that hangs in the air and shrouds the accompanying scenery from view. The old joke goes: *If you can't see the forests, it's raining; and if you can see the forests, it's going to rain.* That wet weather can severely disrupt proceedings on track, a lack of visibility being the primary issue for drivers. That is because of the volume of water kicked up by the Formula 1 cars – accentuated by ground effect regulations and the wide tyres. On damp and misty days the spray tends to hang in the air through the forests, which means the track can be grippy enough to be raceable, ensuring drivers aren't aquaplaning haplessly into the scenery, but racing cannot happen due to the visibility and the risk of striking a rival unsighted.

Says Mercedes' George Russell: 'The way I describe it to try and give some perspective is driving down the motorway in pouring rain and turning your windscreen wipers off. That's genuinely how it feels in the cockpit.'

Belgium's 1998 race was held in torrential conditions and included the largest first-lap incident in history, when a dozen cars were caught up in an accident exiting the La Source hairpin. Fortunately, there were no serious injuries and the race went on to be won by Damon Hill, marking the first grand prix win for Jordan, the team which morphed through various guises into Aston Martin. In 2021 Spa-Francorchamps was shrouded in relentless rain and mist, and conditions were deemed unsafe to race. Only three laps of a race were completed behind the safety car before the race was suspended and not resumed. Due to Formula 1's rules the results were backdated to the end of the first lap, meaning the race officially lasted just 3 minutes, 27 seconds.

It is an event that has always attracted a hardcore fan base, who come prepped for any and all weather conditions. In the 1990s and early 2000s, this was Schumacher territory. Michael Schumacher is synonymous with Spa-Francorchamps, having made his debut at the circuit in 1991, then going on to take a maiden victory a year later, and in 2004 he clinched his seventh and final world title at the venue. It was also something of a home event for Schumacher, given that he hailed from Kerpen, across the border in Germany but only around 80km (50 miles) away. In the last decade the race has been increasingly taken over by the Dutch courtesy of the Verstappen effect. And while he is Dutch, and competes under a Dutch licence, Verstappen's life began in Belgium.

'I was born in Belgium,' said Verstappen. 'I grew up in Belgium, just across the border to go to school. But also after school, I would travel just across the border to Holland, where the workshop was, with all the go-karts and stuff. So yeah, you know, it's a big part, my Belgian side of the family as well. And I also really enjoy it.'

The passionate spectators spend their days perched on steep grassy spectator banks and their nights in gradually deteriorating campsites, nestled in the fields and forests that straddle the contours of the expansive circuit. Some of those fields are used for parking, so the presence of rain turns them into quagmires and the police convey little interest in assisting. There is an attempt to execute a traffic management plan, but it habitually collapses, exacerbated by illogical one-way systems that create unnecessarily lengthy diversions, and hamstrung in the first instance by the region's fiddly and pothole-laden roads. Knowledge of the back roads and shortcuts are useful only until encountering the resistance of a stern-faced police officer and their unfriendly canine partners. Police resources can also be stretched during the Belgian GP weekend, because the event now clashes with Tomorrowland, a major music festival, which takes place to the north of Brussels. Access to Spa-Francorchamps has to be by car – or on a bicycle for the very bold – because of the lack of public transport infrastructure in such a rural area. The nearest international airports are Brussels, Luxembourg or Cologne, and while there are options to reach the region by train, buses from Liège or Verviers are unreliable.

Spa-Francorchamps has history and is a classic circuit, but its future remains uncertain. It struggles to increase capacity due to its topography and logistical challenges, while the flow of people both within the circuit and through the surrounding roads may reach a breaking point.

Its recent contracts have just been a spate of short-term extensions, covering events in 2023, 2024 and 2025. When the most recent contract was announced, F1 CEO Stefano Domenicali noted the 'big strides' made in terms of infrastructure and fan entertainment while hinting that more work is required. Nonetheless, it is obvious that Belgium cannot compete with some newer facilities.

'I think the concern is probably there,' says Vandoorne on the event disappearing. 'Formula 1 is looking to expand their calendar more and more, and you can see a lot of the new venues that are coming to Formula 1 are quite commercial venues, like Miami and Las Vegas. They're big places that they're going to, and I think that's fine, absolutely fine, but I still think there needs to be a bit of a combination of both on the calendar, to retain these races, like Spa, like Suzuka, that are the proper old-school tracks.'

Belgium at least still has a race, which is more than can be said for its larger neighbour, France. France is one of the homes of motor sport: it is where governing body the FIA is based and where one of the most famous events in motor sport – the Le Mans 24 Hours – takes place. The first motor-sport events took place in France in the late 19th century while the country gave the world the phrase 'grand prix': the French Grand Prix was the first to take on the moniker, in 1906, 44 years before the creation of the Formula 1 world championship. Fearsome road courses in Reims and Rouen hosted some of Formula 1's early French races, along with Clermont-Ferrand, while Le Mans' Bugatti Circuit made an unpopular one-off appearance. As Formula 1 became increasingly commercial, permanent venues Dijon-Prenois and Paul Ricard held grands prix, during an era in which France had several front runners and race winners, Alain Prost taking four world titles. France's grand prix in 1982 had a home 1-2-3-4, with René Arnoux victorious, ahead of Prost, Didier Pironi and Patrick Tambay, while the leading duo were driving Renault cars, using Michelin tyres – the most French of French results possible.

Formula 1 moved to Magny-Cours in 1991, bang in the centre of France, but the remote race at a so-so circuit was rarely popular and as the 2000s progressed the financial situation became bleaker. France withdrew as a promoter prior to 2009, marking the first year since 1955 in which a grand prix did not take place in the country. Alternative locations and ideas were mooted, but France remained absent for almost a decade.

France returned in 2018, with a renovated Paul Ricard and its garish blue and red striped run-off, revived for a five-year deal, having last hosted Formula 1 in 1990. The first return event was marred by traffic chaos and access issues, with some stuck for hours on the twisty roads that lead up from the coast to the venue on a plateau, and though the situation improved this remained a hard reputational mark for organisers to remove. France remarketed itself as 'La Summer Race', leaning into its high season placement on the schedule, and Paul Ricard's close proximity to the gorgeous sun-drenched Provençal coast with its beautiful towns, world-leading cuisine and rolling lavender-stained fields. But Paul Ricard was never financially successful in the long term: France was not a market that Formula 1 was enormously desperate to exploit and leading politicians displayed little interest in securing a round of the championship beyond the usual lip service. The circuit itself was merely average and the races were not exactly a thrill a minute, and after 2022 Formula 1 bid adieu to France.

'It's not normal that we don't have a French Grand Prix at the moment,' said France's Esteban Ocon in 2023. 'It should be part of the calendar. I hope we will find the solution together to bring it back.'

Only six countries have hosted more grands prix than France's 63, but there has been little indication that Formula 1 will soon return.

15

HUNGARIAN GRAND PRIX: HUNGARORING, BUDAPEST

'I love driving at the Hungaroring... It's a proper driver's circuit with lots of twists and turns, where overtaking can be difficult and tyre management is very important.'

ESTEBAN OCON

'It started probably around 1983, when Bernie Ecclestone decided that it'd be amazing to have a race behind the Iron Curtain,' says Ariane Frank-Meulenbelt, Vice President of the Hungaroring and promoter of the Hungarian Grand Prix.

Formula 1 was increasing in popularity, fuelled by the growth of its TV product, and Ecclestone was exploring options.

'He was flying over Budapest with one of his friends, who happened to be a Hungarian national, and he said, *Why don't we do it in Hungary? I know some places,*' she explains. 'As the legend goes, they looked out of the plane window and said: *What's this area here?*, and Bernie took a piece of paper, drew out something which he thought would work well, they started talking to the Hungarian officials and realised it's a good place to host a grand prix.'

Hungary's grand prix is now approaching its fifth decade of existence. The Hungaroring is surpassed only by Monza as the circuit with the longest unbroken stretch on Formula 1's calendar, the championship having visited for 38 straight years and counting.

Hungary was behind the Iron Curtain, but there was a voracious appetite for motor sport.

'I grew up in a small countryside town in the Eastern part of the country where opportunities were really limited in every single sense,' says Sándor Mészáros, Hungary's sole full-time Formula 1 journalist. 'My late father was technically oriented and an avid fan of motor racing. Hungarian national TV – the only legally available channel – covered Formula 1 from 1976 and increased the coverage, but they only showed selected ones. I had a huge hunger for Formula 1 and my father collected some money and when he had to go to Vienna on business – which was very rare in those times to go across to Western Europe – he smuggled a TV antenna back and that meant we could watch Austria's TV coverage. That was a huge risk, as using an antenna like that was forbidden by the government and you could even lose your job as a punishment. My father hid it under the roof of our house and when we wanted to use it, he moved the tiles.'

The Hungaroring, which received sizeable support from the authorities, was constructed in nine months and opened its doors for its first Formula 1 race in August 1986. An estimated 200,000 spectators flocked to the edge of Budapest – and it proved a seminal moment in Hungary's history.

'My father planned a visit to Austria for 1985 to see F1, but the news came out that we would have a new track near Budapest and a race in 1986, so he decided to wait and skip the trip,' Mészáros recalls. 'From that money he bought the antenna and the next year, on 10 August, 1986, as a six-year-old kid, I sat on his shoulders at the Hungaroring and watched the race at the main straight.

'Even though I was very young, I remember very well this period and I always say that there were three key things that contributed to the changes of that world and the collapse of the Iron Curtain. The first one is the election of Pope Saint John Paul II in October 1978. He came from Poland, right from the heart of the communist bloc, and made a huge influence on all countries behind the Iron Curtain. The second big thing was the mega concert of Queen in Budapest, on 27 July 1986. It was the first time an absolutely world-class rock band came to the communist bloc to play and it was a massive hit. Two weeks later Formula 1 was on track at Hungaroring, the very first F1 race at the very first purpose-built F1 track of the communist bloc. The Western world kicked the door on the Eastern bloc, made a huge impact and changed so many things in the minds of the people who realised, *We are fed up with what we have and we prefer this.* Those vibes, feelings, the winds of change had started and it contributed significantly to the collapse of the Soviet bloc. This is the reason why for me the Hungaroring is not just a racetrack: it's a symbol of the past and the people's wish of freedom.'

As Hungary escaped from the clutches of communism, the Formula 1 grand prix became a focal point of the summer, with each of its races having taken place between late July and mid-August, enabling promoters to tap into the swelteringly sunny days – and stifling nights – of beautiful Budapest.

'Initially the event had a lot of Hungarians, and Russians could come and visit, but it wasn't really open to other nationalities,' says Frank-Meulenbelt. 'Now it's a very international event, with about 70 per cent foreigners.'

The occasion brings together an eclectic range of nationalities throughout the weekend; while Verstappen's Orange Army have increasingly invaded the Hungaroring, the central European location means it is attractive for a breadth of nationalities and there has often been a large Scandinavian contingent. An army of yellow taxis zips spectators to and from downtown Budapest, and there are always huge crowds cramming around the crossroads at the circuit gates.

The Hungaroring itself is a fiddly and technical circuit, where overtaking is tricky. It is one of the shortest on the schedule, there is only one straight of note – the main straight – and most of the corners are medium-speed turns around narrow sections of tarmac.

The Hungaroring has nonetheless proffered some exciting moments throughout Formula 1 history, including the location of maiden grand prix wins for the likes of Fernando Alonso (2003), Jenson Button (2006) and Esteban Ocon (2021), who won after two separate first-corner pile-ups wiped out most of the front runners.

'I love driving at the Hungaroring,' says Ocon. 'It's up there as one of my favourites and that's not just because I won there! It's a proper driver's circuit with lots of twists and turns, where overtaking can be difficult and tyre management is very important.'

Hungary has only had one Formula 1 driver in history – Zsolt Baumgartner, who started 20 grands prix across 2003/04 and debuted at the Hungaroring – but motor sport is big in the country.

'People don't realise we've got quite a lot of automotive industry in the country. Mercedes and Audi build a lot of their cars in Hungary,' says Frank-Meulenbelt. 'There's a lot of people in junior levels, the very few make it to the high end, but there's a vibrant scene – rallying, rallycross, go-karting – in Hungary.'

Hungary will always have a place in the history of motor sport. The winner of the first grand prix, the French Grand Prix in 1906, was the Hungarian Ferenc Szisz. He is commemorated with both a statue at the main entrance of the Hungaroring and a plaque marking the feat in the lobby of the pit building. Hungary also held a grand prix in 1936, when the likes of Mercedes and Auto Union raced through the avenues of Budapest's Népliget (People's Park).

While some European venues face uncertain long-term futures, Hungary's presence on Formula 1's calendar will stretch onwards with a contract to host a grand prix until at least 2032. The facilities are outdated and the infrastructure has a very 1980s Soviet feel, but a multi-year renovation project got underway in 2024.

'Our circuit is nearly 40 years old and there's been a few changes but nothing majorly,' says Frank-Meulenbelt. 'What we are doing now is bringing the infrastructure into the modern era: a whole new pit building, a lot of new grandstands, a new event area behind the main grandstand. We want to make the experience better for our fans and second of all to bring our facilities up to the standard that F1 now requires – our garages are very small, we need to expand the space and to make the environment more comfortable for them.'

Looking to the next decade while preserving the Hungaroring's heritage is a key ambition.

'The management of the Hungaroring, the promoter and the government are thinking on an even longer term,' says Mészáros.

'The current investment is a strong sign that their intention is very serious. I have no doubt that we are on the right track. Due to the historic ties, F1 fanaticism is coded in the Hungarian people's DNA. For the generation of my parents, F1 was a symbol of the new world, for me it was a dream, while the generations after me grew up with F1 passion as a part of their everyday life.'

As Formula 1 approaches the middle of the season, the driver market for the following year begins to click into place – though increasingly that is happening earlier in the year as teams try and lock down drivers for the long term. But teams lower down the grid tend to wait a little longer and that also means attention is increasingly paid to the next generation of racers.

Formula 1 is supported at just over half of the grands prix by Formula 2 and at about a third of the events by Formula 3, the categories in which young drivers race in order to develop their skills while proving themselves to the bosses in the big paddock.

Formula 2 evolved in 2017 from the GP2 Series – which replaced Formula 3000 in 2005 – while Formula 3 was reborn in 2019 after a merger between the GP3 Series and European Formula 3.

'Formula 2 and Formula 3 are part of what is called the single-seater pyramid, more or less starting from go-karting,' says Bruno Michel, the long-time CEO of Formula Motorsport, which runs both championships.

'It's a pyramid, so the bottom has quite a large base, then it gets thinner. It's the way it's working, as the following step after Formula 2 is Formula 1. We've always made sure that the main goal of Formula 3 and 2 is to prepare drivers to go to Formula 1 – not all of them will go there, of course.'

Each team in both championships receives the same single-spec chassis, which usually lasts for six years before being updated, and it is up to the mechanics and drivers to extract the maximum from their packages. There is still a hierarchy in terms of the standout teams, but they are categories in which any driver can triumph, or at least deliver standout performances and capture the attention of Formula 1 teams.

Formula 2 cars tend to be at around 115 per cent of the lap times of Formula 1 cars, with Formula 3 about 120 per cent.

'We put together a championship that is quite clear in terms of what we want to achieve,' says Michel on Formula 2 and 3. 'Both categories are single-make categories and that means everyone has the same car, and if you give the same car to everybody – of course teams might have differences in the way they prepare the car or prepare the drivers – but at the end of the day they are drivers' championships, which means the best drivers win and the best drivers go up, and it's an *up or out* system, to make sure they can reach Formula 1. We have to make sure the car is going to be at the right level of performance to make sure Formula 3 drivers are prepared for Formula 2, and it's a step, and Formula 2 drivers are well prepared for Formula 1. If you look at the drivers who went to Formula 1, they've always been ready when they get there. It means the car they've been driving has been at a proper level and they're driving during Formula 1 weekends.'

The Formula 2 car has to be technologically advanced in order to best prepare drivers for modern Formula 1 machinery. In fact, Formula 2 is sometimes used as a test bed for the bigger championship, as has happened in the early 2020s with wider tyres and synthetic fuels.

That creates a problematic trade-off: how to keep down costs. The championships partly alleviate the problem by limiting the number of permitted personnel, with 12 people allowed per two-car F2 team and only 10 for a three-car F3 team, and restrictions on official testing. But teams still need to find the required money to run, maintain and service the cars – while employing those personnel – and so it is up to the drivers to find the budget. For a front-running Formula 3 team, that is now over $1 million for a season, and for Formula 2 that figure can easily be doubled, tripled, or more. This is becoming an increasing problem, particularly at entry level and even as low down as karting. Anyone in the world can kick a ball, but to go racing you need not only access and facilities – a racetrack, karting facility and equipment – but money.

'Arriving to Formula 1 is expensive and the drivers have to find the money. Some of them are funded, more and more are belonging to Formula 1 team academies, which makes things easier, but sometimes the academies do not fund the whole amount, so then it partly comes down to the driver, and sponsors, having separate funding to make sure they can access Formula 1,' says Michel.

'It's a question everyone is asking me, *Are we sure that all the best drivers are going to Formula 1?* And I've been doing this job more than 20 years, and I don't think I have one example of a driver who was really outstandingly good and who didn't make it.'

Formula 1 organisations have increasingly cottoned on to the notion of having young driver programmes. These have existed in various functions throughout Formula 1, but have become more formalised since the turn of the century. Red Bull's scheme revolutionised the landscape by creating a direct Formula 1 pathway via its controversial second Formula 1 team.

Red Bull, Mercedes, Ferrari, McLaren, Alpine, Sauber and Williams all have young driver academies, while power unit supplier Honda also has its own scheme. Latching on to a scheme can work both ways: it helps drivers navigate their way to Formula 1 – and at cut-price – while also giving Formula 1 teams the opportunity to secure prospective young talents with long-term arrangements. The likes of Charles Leclerc (Ferrari), Lando Norris (McLaren) and George Russell (Mercedes) are just some of the high-profile names who were signed by their respective team academies prior to starting out at Formula 3 level.

'There is a lot of work going on behind the scenes,' says Sven Smeets, Sporting Director at Williams. 'The drivers are helped mentally, physically, with media training, social media training, we can bring them into a culture, they can talk to engineers, do simulator programmes in Formula 3, 2, 1. We want them to go to school and not miss out on this, as education is important, and the social contact, so all this goes into the programme and it's fantastic to work with them – they all have this dream to be a Formula 1 driver. Unfortunately, it's not always possible. It's not like the Premier League with 20 teams and 11 players, there's a lot of work behind the scenes, we've seen talents not make it. It is a very hard journey.'

Junior drivers often gradually integrate themselves within Formula 1 teams at grand prix weekends as they progress through the categories and take on more responsibilities within the team, such as simulator or reserve driver duties. That can mean extensive running in a virtual world behind the scenes, assessing suggested tweaks between real-life practice sessions and honing a set-up direction, while also being on hand if a team's regular driver is unable to compete. Attitude and application in such circumstances can also facilitate a promotion.

Oscar Piastri, who had to leave his native Australia aged 14 to pursue his Formula 1 dream, secured successive titles in Formula 3 (2020) and Formula 2 (2021), when he was part of Alpine's Academy, an association that still required him to bring personal funding. Piastri was infamously part of a tug of war in mid-2022, when he jumped from Alpine – where he was its reserve driver – to McLaren to secure a 2023 Formula 1 seat. The Contract Recognition Board ruled in McLaren's favour.

'The pressure to perform is quite similar for differing reasons,' says Piastri. 'In Formula 1 there's more media attention, there's more spotlight from fans, and I guess there's the external pressure on you, and of course pressure to keep your seat. But the pressure to even get to Formula 1 in the first place, and how difficult it is, is very, very tough as well. It's either family-funded or sponsor-funded. So there's that sort of financial pressure in a way. There's a pressure to get through the categories quickly and kind of ride the momentum and stay in the spotlight. So it's very tough going up the ladder in some ways. That, I personally think, prepares you well for Formula 1.'

It is nevertheless an elitist sport and almost impossibly hard for a driver coming from a conventional background. Youngsters from wealthy backgrounds, or who have a benefactor, have an ingrained advantage and there have been several cases of drivers with limited talent – still competent but hardly outstanding – buying their way to the top. Those with pure talent who do make it to the top of junior motor sport have traversed the narrow pyramid, myriad hurdles and setbacks, but still must secure one of only 20 seats at the very top table.

Formula 1 was joined in 2024 by F1 Academy, also run by Formula Motorsport, which is a Formula 4-level series only for female drivers. The absence of female drivers from Formula 1 has been a talking point for decades and former Williams test driver

Susie Wolff is a key figure at the helm of F1 Academy, driving the quest for change. The category was launched with the backing of Formula 1 in 2023, a year after the collapse of W Series, which ran for three seasons and opened the door for aspiring racers before encountering financial problems. F1 Academy's aim is to prepare female drivers for the next steps of the FIA's single-seater pyramid, while also providing support on physical, mental and technological development. It is not merely about what the existing crop of drivers can achieve, but also about providing awareness and opportunities for the next generation – behind the wheel, as well as across a range of employment openings in motor sport. The hope is that this will lead to greater involvement at a grassroots level and therefore a significantly wider talent pool. The success of F1 Academy – and the legacy of W Series – will only be truly measurable into the 2030s and beyond, but proactive steps are at least being taken.

Budapest straddles the Danube, with the Hungaroring just outside of the city

16

SUMMER BREAK

'It's probably one of the best things we ever introduced in Formula 1. This is really the only time in the whole year where you can relax and you can switch off...'
BEAT ZEHNDER, SAUBER TEAM MANAGER

Between March and July Formula 1 has got through over a dozen grands prix, with the mid-year months regularly featuring a spree of events.

In the middle of the Formula 1 season comes the summer break. As the calendar has expanded, with more events added, a four-week gap has been introduced between consecutive events in late July and late August. Given the aftermath of the final pre-summer race, and the build-up to the first post-summer race, that realistically makes it a three-week gap, but a firm shutdown is written into the regulations to cover a 14-day period. As with anything Formula 1, there are strict and detailed regulations: Formula 1 teams are not allowed to design, produce or develop any car components during those 14 days, effectively meaning that the sport shuts up shop for an extended break.

It means personnel have two weeks free – from work, from emails, from meetings and from having to think about Formula 1 (or colleagues). Each team can choose when to implement the 14-day period during the gap between races and must notify the FIA of those dates. The shutdown also applies to power unit manufacturers. Teams must take the 14-day shutdown into account when planning for the long term and can rest easy in the knowledge that their nine competitors are also not working on car performance. The summer break was joined in 2023/24 by a nine-day shutdown that covers December 24 to January 1 inclusive.

'I think the summer break is a really great thing for everybody in the sport,' says George Russell. 'We are all so motivated and determined and everybody in their own right feels like a bit of a warrior and just wants to power through. But I think everyone recognises the impact that two-week break has to totally reset, come back rejuvenated and also come up with some new ideas.'

For team members the shutdown is often the only moment to truly decompress, because everyone knows that everyone else is in the same predicament.

'They do what they say on the tin and the business and the sport shuts down,' says Pete Crolla, Haas F1 Team Trackside Operations Manager. 'Nobody works in that time because they can't. And those, certainly from an operational perspective, they can't work. There are some supporting elements of the business that have to continue, but they know to leave you alone. So for two weeks in summer and a week in the winter, the phone doesn't ring, there's no emails coming through. And that is genuinely a time when you can shut down because you can take holiday during certain points of the year, but the rest of the business is still operating. So you're still getting phone calls, you're still getting emails and you can't truly switch off from it because modern society and modern business doesn't respect time off. I think that's similar for a lot of industries now. And even if you're on holiday, if somebody needs something now, everybody is too contactable. You need these moments where the entire business shuts down to actually shut yourself down.'

It also provides an opportunity for the paddock to refresh after the opening half of the year.

'You see that there is a sort of degradation that you have towards the season up to August,' says Marco Perrone, Sporting Director at RB. 'And then you see everybody's fresh, reset.'

Says Beat Zehnder, Sauber Team Manager: 'It's probably one of the best things we ever introduced in Formula 1. This is really the only time in the whole year where you can relax and you can switch off because we work on race-free weekends. The number of emails I get on the non-race weekend is insane: whether it's with the FIA or the first groups that are travelling the weekend before, such as the setup team and the catering team. Every now and then you have problems, because maybe the flight didn't go, or the rental car isn't ready, so you're consistently under pressure. In these two weeks we have nothing.'

It is not entirely a case of the lights being switched off for 14 days and facilities remaining dormant. The regulations list exceptions that mean marketing, financial and legal departments remain operational, while teams are permitted to invest in infrastructure or carry out operations related to the running of historic machinery or demonstration events. One crucial element is the ability to carry out maintenance of equipment or a deep clean of the factory given that there aren't as many people in the way and those that are present are not utilising the machinery required for car development.

For drivers this is a golden opportunity to switch off for a couple of weeks, visit family and friends, and get away from the grind of the schedule.

'I really treasure both the winter and summer breaks,' says Pierre Gasly. 'It's so important to recoup energy in order to stay fresh and maintain high levels of performance, which is required from us at the racetrack.'

However, it isn't a complete reset, because there's half of the season left to run.

'Quite often I like to travel somewhere, in a different time zone and everything. I do try and switch off,' explains Valtteri Bottas. 'But in the end, you never switch off fully. Your phone is on, every now and then you might get something come to your mind and you want to tell the engineers about it and vice versa. You can't switch off 100 per cent, as you know there's still half of the season to go – but even semi-switching off is fine.'

In spite of – or perhaps because of – the summer break, there can be movements in the driver market.

Formula 1's 'silly season', that time of the year given over to rumours and conjectures about who is going where, has its fallow cycles and then some years where the marketplace explodes into life, increasingly taking place earlier in the year. Frequently there are one or two major pieces of the puzzle – typically the leading drivers at top teams – and that triggers a frenzied domino effect as various representatives, team bosses, drivers and journalists try to get an idea of the lie of the land.

Those with cast-iron contracts can relax in the knowledge of future employment while those with expiring deals – or eyeing a move to what they think is a better team – start to get a wiggle on. Sometimes teams or drivers will move to activate a retainer or exit clause, depending on their intentions, before a certain deadline, and teams have been known to pay off a driver in the event of underperformance amid the belief that the grass is greener somewhere else. There may be handshakes and the signings of pre-contracts, but no follow-through. On very rare occasions the FIA's Contract Recognition Board must intervene to settle disputes and legally declare the validity of who has signed what with whom and when.

The market also extends to senior personnel, and the top engineers and technical figures are headhunted, often with a view to the long-term future. Key figures have contracts that require a period of 'gardening leave', potentially as lengthy as 12 months, meaning that unless an agreement is reached between respective parties they must wait for a prolonged period before swiping into their new office.

17

DUTCH GRAND PRIX: CIRCUIT ZANDVOORT, ZANDVOORT

> *'I like the track, it's old school and unique, and it's one of those that you definitely enjoy driving a Formula 1 car on.'*
> KEVIN MAGNUSSEN

The Dutch Grand Prix was part of Formula 1's calendar from 1952 to 1985, with the fast and flowing Circuit Zandvoort, located in the sand dunes adjacent to the North Sea, its permanent home.

Racing first came to Zandvoort in 1939, when a competition was arranged through the town's streets and plans to construct a permanent facility were tentatively drawn up. Those with even a rudimentary knowledge of history will realise that summer 1939 proved not to be a good period in which to make long-term business plans in Europe. And yet, weirdly . . .

Zandvoort as a town suffered sizeable damage during the early years of the Second World War: most of its buildings were levelled and swathes of reconstruction were required.

The legend goes that Zandvoort's mayor, Henri van Alphen, continued in secret with plans for a racetrack and convinced the Nazi occupiers that they should construct a lengthy wide piece of road in which to hold a parade after their presumed victory. Plot spoiler: such a scenario never came to fruition and instead Zandvoort had the outline of a pit straight, as well as several undulating routes that had been constructed as communication roads. Motor sport was not exactly a priority in

1945, but once the region had sufficiently recovered, attention turned to the rubble to the north of the town and the layout was paved. The circuit opened for use in 1948 and four years later Formula 1's world championship set up its Dutch Grand Prix at Zandvoort.

As the 1980s developed and Formula 1 grew as a business that increasingly chased higher revenues, the old-fashioned Zandvoort dropped from the schedule. Land was sold off to property developers and part of the circuit was lost forever, with a truncated yet still challenging layout eventually revived in 1999 to play host to national championships. Single-seater machinery continued to grace Zandvoort's tarmac, its pinnacle event the Masters of Formula 3, won in 2005 by Lewis Hamilton and in 2014 by a young local chap called Max Verstappen.

It was the emergence of Verstappen that ultimately triggered Formula 1's return to the Netherlands.

Formula 1 has always been popular in the Netherlands, with both motor sport and motorcycle racing attracting enthusiastic crowds, and there has been passionate support for home-grown talents. That was most obvious in the mid-1990s, when a hardcore group followed Jos Verstappen, whose career peaked with two podium finishes. A smattering of Dutch drivers followed, including Christijan Albers, Robert Doornbos and Giedo van der Garde, all of whom reached a ceiling towards the back of the grid. But through the late 2000s there was already a focus on Jos' son, Max, who was ripping up the scene in karting, guided by his father's strict hand.

Verstappen began car racing in 2014, leaping straight from karting to Formula 3, a decision that gained widespread attention as he bypassed the usual entry categories in junior motor sport, such as Formula 4 or Formula Regional. He was immediately a front runner and by midseason, still aged only 16, had piqued the interest of Formula 1 teams. Mercedes was interested in securing his services, but so too were Red Bull. They swooped mid 2014 with the offer of a Formula 1 seat for 2015, at Toro Rosso, and Verstappen made his debut at 17 as the youngest racer in history. He was on the grid, at the elite level, just 15 months after signing off from karting, having also skipped Formula 2 (which was at the time called GP2). After 23 grands prix he was promoted to Red

Bull's senior team and on his first start, at Barcelona in 2016, he claimed victory. Aged only 18, Verstappen became Formula 1's youngest race winner, shattering the previous record by two-and-a-half years and capping an unprecedented rise.

'I think if there was something that Max never lacked, it was hype around him,' says Carlos Sainz, Verstappen's first teammate in Formula 1, in 2015. 'I think he had the hype even well before his debut in Formula 1; when he was already in F3, he had already a lot of following. Obviously he's a special talent and someone that is performing really, really well since the beginning.'

Pierre Gasly was teammate to Verstappen at Red Bull in 2019. 'It was clear already from the very young age when we raced in karting, the hype around him was already very different to any other drivers,' he says. 'It was clear going up the ranks that there was more attention, more focus, driven by the name, the performances, all the hype around Jos and Max's relationship. So, it was no surprise, and the way that he came to Formula 1 and performed and switched to Red Bull just amplified everything that followed. It's no surprise that he's where he is today and he's got the following, especially in the Netherlands. From an outside point of view, he seems to be pretty much the icon of the country. So it explains why 99.9 per cent of the grandstands are full of orange when we come to Zandvoort!'

Verstappen's swift rise and emergence as a Formula 1 front runner led to a boom in Dutch spectators at a plethora of European grands prix, most notably across the border in Belgium, and at Red Bull's home event, in Austria. The grandstands became increasingly coloured by the orange of the Netherlands, there were specific areas for Verstappen's fans and decorated caravans pledging support to the rising star traipsed around the highways of Europe. In Verstappen's home country there were even tentative discussions over whether the Dutch Grand Prix could be resurrected.

'I remember in 2017 there was a press conference at the track and they had a big survey and investigations as to if it was ever possible to get Formula 1 back,' says Dutch journalist Erik van Haren from daily *De Telegraaf*. 'They said it was possible, but I think everyone in the room thought it would never happen – it was not a complete plan, it was, *OK, it is possible*.'

Under Formula 1's relatively new owners, Liberty Media, the prospect of a city race was briefly explored, but it was rapidly ruled out. Neither Amsterdam nor Rotterdam were willing to accept the disruption caused by a grand prix, nor were they truly realistic locations. That left the permanent circuits as the only viable candidates: Assen, a venue in the north-east of the country intrinsically associated with motorcycle racing, and coastal Zandvoort. It was abundantly clear that Zandvoort was the front runner and it secured an exclusive agreement to discuss a deal with Formula 1 bosses. By early 2019 the relevant assessments had been completed and a contract was signed for the Dutch GP to open the European season in May 2020. Here comes a second plot spoiler . . .

The Covid-19 pandemic meant the event was cancelled for 2020, given that Zandvoort was going to be an *Oranje* festival, heavily reliant financially upon ticket sales. A revised calendar placed the Netherlands between Belgium and Italy on the 2021 schedule, in late summer, and when Formula 1 finally returned to Zandvoort it was clearly a case of better late than never. The crowd got what they came for: a victory for Max Verstappen, en route to his maiden world title.

'It's a unique atmosphere, the promoters do a great job,' says Red Bull Team Principal Christian Horner. 'Arguably, it's too small for Formula 1, but then so is Monaco. The race is on the calendar for one reason, which is Max, and the effect that he's had. When you come here and feel the atmosphere it's a phenomenal event.'

As a circuit it is popular among the drivers: twisty, old-school and with little room for errors. The first half of the track undulates across the dunes, including the high-speed downhill plunge at the almost-blind Scheivlak curve, while the newer second half of the circuit is flatter but deceptively tricky. Ahead of Formula 1's return two corners were reprofiled in order to add a distinct challenge: banking. At the medium-speed hairpin Hugenholtzbocht – a turn named after the circuit's first director, Hans Hugenholtz – progressive banking of up to 18 degrees was added, creating a bowl-like turn in front of the walkway between Zandvoort's split paddock set-up. The final corner, Arie Luyendykbocht, was also transformed: a long-radius right-hander became a sweeping curve that features banking of 15–18 degrees, propelling drivers on to the pit straight at a higher speed.

'I like the track, it's old school and unique, and it's one of those that you definitely enjoy driving a Formula 1 car on,' says Kevin Magnussen.

Zandvoort itself is a charmingly twee coastal resort, pleasantly warm and welcoming in the late summer sun, though prone to days when the impending autumnal chill strikes early, sweeping blustery winds across the region. Its beach stretches out for miles and in the summer there are a variety of diverse seasonal restaurants, cafés and clubs laid out on the sand, attracting racegoers and holidaymakers alike. The lengthy bricked pedestrianised promenade acts as a barrier between Zandvoort and the sea, carving its way alongside the functional low-rise, sea-facing apartment blocks, snack vans (one is, imaginatively, called Snackvoort) and chip stalls.

Zandvoort is a holiday destination and has its own Center Parcs, which was partially built across the old circuit layout and is taken over for the Formula 1 weekend by half of the paddock – marking an unusual residency for a grand prix.

The Dutch Grand Prix is comfortably the biggest event for Zandvoort, which has a population of only around 20,000, making it one of the smallest communities visited by Formula 1. That raised obvious concerns about the local infrastructure and capacity in advance of Formula 1's return. How do you get approximately five times the population of the town into and out of the area, each day, without disruption?

The Dutch have meticulous plans and execute them to near-perfection.

The few small roads in and around Zandvoort are effectively closed to everyone except the town's residents and certain Formula 1 passholders throughout the course of the weekend, meaning that spectators have to seek alternative means of getting there.

The Dutch inclination for cycling is also fully exploited, with one-way systems in place in the region and spectators urged to travel in that way, leaving their bicycles in the plethora of cycle parking areas. An estimated one-third of spectators reach Zandvoort on two wheels.

Another method of mass transportation is to utilise the Netherlands' effective railway network. The information screens at Amsterdam Centraal helpfully have a racing car symbol next to all the Zandvoort-bound trains and these are affectionately nicknamed 'The Max Express'.

'It's amazing what they do. When you go to Zandvoort on a non-F1 race weekend you wonder how on earth it is possible to have a Formula 1 race here: it's an old place, there's two roads to the track, one from the south, one from the other side,' says van Haren. 'It's amazing what they do. I never thought it would have happened and a lot of people were quite *OK, let's see*. Everyone was criticising them, as well me, and we were like *OK, let's see, they said it's possible*, and they proved it's possible. The first year [back] in '21 I thought: *Wow*, and we had great weather, the atmosphere was great, people were using the train and the bicycles, it's a unique event, the track is unique, I like to go there. It's a mental week for me, but when you're there it's very special. I have to enjoy it because you never know how long they will be on the calendar.'

The event remains a festival of Verstappen. The grandstands are full of fans clad in his orange merchandise – shirts, caps and capes – while flags flutter and the occasional orange flare is launched before dissipating through the often windswept air. Techno music is relentless, with regular chants and songs about Verstappen, including the regularly played 'Super Max' and '33 Max Verstappen', which have evolved into football-style anthems.

'Without Max it was not possible,' says van Haren. 'And it never would have been. Of course, F1 without Max still has a fan base in the Netherlands, as it had in the past when there weren't any Dutch drivers, or just one at the back of the grid, but like this, for such an individual sportsman there is support I never saw before. Without him there's not a chance we'd be in Zandvoort.'

Verstappen's presence and success has brought Formula 1's popularity in the Netherlands to second place, behind only football: bars and pubs hold events to show grands prix, a dynamic that was unthinkable in a pre-Verstappen era.

'I think it's just great,' said Verstappen in 2023 on the event. 'It doesn't bring a weight on my shoulders of extra pressure. I think it's just amazing that this is possible. I think nobody like 10 years ago even thought about a grand prix here, and that we're able to do that now is just fantastic and hopefully it will continue for a while.

For me it's amazing to be here, to see all the fans and drive such an incredible track.'

It has so far been a very happy hunting ground for Verstappen, head and shoulders clear as the most successful Dutch driver in history, taking victory in 2021, 2022 and 2023.

There is the inevitable question of the long term. Zandvoort has never had a lengthy race deal, compared to some of the more lucrative events on the schedule, and despite its popularity it is not a financially rewarding grand prix for organisers. It is a rarity in receiving only minimal local funding and none on a national level. There is limited scope for increasing the attendance of approximately 100,000 spectators a day due to the confines of the small venue, while the bigger aspect is Verstappen. He will not be around forever, so what happens to the Dutch Grand Prix when he exits Formula 1? There is every chance an annual Dutch Grand Prix could be a short-lived feature, but while it lasts, and while Verstappen is the phenomenon he is, everyone present is determined to soak up every last minute.

Max Verstappen, tackling Zandvoort's banking,
is responsible for the Dutch GP's resurrection

18

ITALIAN GRAND PRIX: AUTODROMO NAZIONALE MONZA, MILAN

> '*There's not many corners, you need to be incredibly precise and you need to drive with nearly no wing. It's very high speed and very high corner speed . . . It feels like ages pass between the corners.*'
>
> DAVIDE VALSECCHI, 2012 GP2 CHAMPION TURNED COMMENTATOR

Zandvoort and Monza are both old-school venues, but the shift in atmosphere, setting and organisation between the events provides quite the contrast.

In the space of just a couple of days Formula 1 departs the expansive windswept North Sea coast and arrives inside the verdant, sun-kissed royal park on the outskirts of fashion capital Milan. The creaking and almost mystical Autodromo Nazionale Monza is the long-standing host of the Italian Grand Prix.

On a clear day, and from the right vantage point on the long pit straight, the edge of the Swiss Alps that perch above Lake Como are visible one way, while the skyline of Milan peeks through from the other direction. And the orange that colours Zandvoort is replaced by the scarlet red of the frenzied *Tifosi*, the name given to the supporters of the team idolised at Monza and throughout Italy: Ferrari.

Ferrari has been part of Formula 1's world championship for the longest period of any team, and is comfortably the sport's most decorated outfit and its most recognisable. It has taken 16 constructors' championships, almost double that of its nearest rival, while its representatives have collected 15 drivers' championships. Its most successful streak came in the early 2000s, when it won six successive constructors' crowns, and Michael Schumacher five drivers' titles on the bounce. It has the most victories, most pole positions, most podiums, most laps led and most 1-2 finishes of any Formula 1 team. The last 15 years may have been leaner for the Scuderia, but it still holds a commanding position atop the record books.

Ferrari is akin to a deity in Italy, its team followed and supported more even than Italian drivers – who have been few and far between lately – and Monza is a place of pilgrimage for its sporting worshippers. There are column inches dedicated to Ferrari's activities in Italy's daily newspapers, sometimes positive, sometimes negative, owing to the political machinations that make Ferrari a team unlike any other on the grid. As Sebastian Vettel once remarked, everybody is a Ferrari fan – even if they say they're not, they're a Ferrari fan.

Ferrari's Formula 1 team has long been based in Maranello, about a two-hour drive from Monza, in a town that is effectively a shrine to the Prancing Horse. Behind its grand racing factory, located on Via Enzo Ferrari, is the Fiorano facility, a private test track where its Formula 1 cars take their first laps in February and where days upon days of testing took place during the era of unregulated in-season activity. The circuit surrounds the white converted farmhouse, with its bright red front door and window shutters, in which founder Enzo Ferrari lived. It has become tradition that fans pack on to the pavement on the SP3 road that runs adjacent to Fiorano, taking advantage of the road's elevated trajectory that rises above the fence, to catch a glimpse of the new machinery that will carry their hopes for the rest of the year. There are usually more fans present for Ferrari's shakedown than for practice days at some grands prix, and it marks the emergence from the dark winter months for the team and its fanatics alike.

The support from the *Tifosi* remains unbridled irrespective of Ferrari's on-track performance. Its drivers are mobbed wherever they go, and serenaded from the roadside on the short drive from Monza's circuit to the exquisite Hotel de la Ville, outside which fans gather several deep for photos, autographs or even just a glimpse of the drivers in red. Somehow pockets of fans even have sufficient intel on where to wait outside the restaurants that Ferrari's drivers and their respective crew frequent.

'We are lucky enough that driving for Ferrari we've got support everywhere, but when we come to Italy, obviously it's on a different level,' said Charles Leclerc in 2023.

The fans in the grandstands erupt each time a red car leaves the garage during practice, the frenzy whipped up by the excitable, 20-words-a-second circuit commentator, and while the drivers are focused on the job at hand, there is usually a cursory out-lap appreciation of those spectators.

Ferrari has won on 20 occasions at Monza, the most wins by a single team at one circuit, and there have been some memorable triumphs.

In 1988 McLaren was chasing a perfect unbeaten season, but an engine failure for Alain Prost followed by Ayrton Senna colliding with the lapped Jean-Louis Schlesser in the closing laps opened the door for Ferrari. Gerhard Berger led Michele Alboreto for a Ferrari 1-2 mere weeks after the death of company founder Enzo Ferrari, inflicting the only defeat that season upon McLaren.

Michael Schumacher, Ferrari's most decorated representative, claimed five Monza wins, bookending his career in red with triumphs at the circuit in 1996 and 2006. After his last Monza victory, in 2006, he and Team Principal Jean Todt emotionally embraced on the podium. Shortly afterwards he revealed his intention to retire from Formula 1, bringing down the curtain on a glittering era.

Schumacher's win count was matched by Lewis Hamilton in the 2010s, while representing Mercedes, a team to which he remained loyal until early in 2024, when he pulled a bombshell announcement that he would be ending the longest driver-team partnership in history to join Ferrari from 2025, a move he described as 'fulfilling a childhood dream [of] driving in Ferrari red.' The Scuderia still has an appeal unlike any other team on the grid.

Speaking in 2023, when racing for Mercedes, Hamilton said: 'I remember there was a period of time where we couldn't wear the team shirt into the circuit, to now just having such a warm welcome from the Italian fans! And finally, for the last, God knows how many years, just feeling really welcome here. It's a place that I've loved since I was like 13, the first time I started racing in Italy, in Parma and Jesolo and South Garda. And it brought me so much happiness. I had so many great memories out here . . .'

Monza, the oldest purpose-built facility in mainland Europe, is now over a century old and has held more grands prix – 74 as of 2024 – than any other circuit since the inception of the world championship in 1950. The circuit is located entirely within the walls of the Parco di Monza, which also houses the neoclassical Villa Reale once inhabited by Italian royalty, featuring swathes of immaculately kept lawns, acres of forest and a golf club. The racetrack is surrounded by the forest, but it was shorn of some of its identity in July 2023 when storms ravaged the region, felling 10,000 trees, some of them crashing down on to the racetrack. A regenerative programme was swiftly launched, but it will take time before Monza's luscious forests, usually starting to be coloured by the oncoming autumn during Formula 1's race weekend, are restored to their former glory.

Monza has missed only one season, in 1980, when the facility was temporarily closed for renovation works, yet the relics of the past remain. Monza was unusual in having a grand prix course as well as a banked oval, and for four grands prix in the late 1950s and early 1960s the two layouts were combined to create one track. The oval was a fearsome challenge, almost vertical at its highest point, and it is scarcely believable that it was once tackled by grand prix machinery. The banked curves, which plunge the drivers between the forests, had just a simple Armco barrier between the track edge and an enormous drop. The banking narrowly escaped demolition in the 1990s and the delicate structure is now preserved, its rising curves perching adjacent to the current first chicane, then crossing the existing layout along the Serraglio straight. It is a regular place of pilgrimage for those in the paddock, to appreciate the history and the ghosts of the past, as well as to see if it's possible to walk up the steepest part of the banking – not to mention the subsequent often less-than-graceful walk down.

Monza is known for one thing: speed. Yes, any motor sport is fast, but at Monza it is all about the straight-line speed, pure power and skinny rear wings.

'It's the *temple di Veloccita*, the Temple of Speed,' says Davide Valsecchi, a TV pundit, who won the GP2 Series title in 2012.

'It's an iconic circuit and has made also the history of Formula 1 – we are fanatical about Monza, because it is a circuit that has a lot of character. There's not many corners, you need to be incredibly precise and you need to drive with nearly no wing. It's very high speed and very high corner speed, with low downforce. It feels like ages pass between the corners.'

Lewis Hamilton completed the fastest ever Formula 1 lap, at Monza, setting pole position in 2020 at an average speed of 264.363km/h (164.26 miles).

Valsecchi, who lives half an hour north of Monza near Lake Como, recalls making the trip to the circuit each year as a child and 'having my mind blown' by Formula 1.

'There was a period when during the August period – before the [now regulation] summer break – there were tests at Monza before the grand prix weekend,' Valsecchi says. 'We were watching there for free, and I remember trying to get into the paddock and to meet the drivers; it was special.'

It is a circuit where surprise results are feasible. Sebastian Vettel famously captured his and Toro Rosso's maiden victory during a rain-affected weekend in 2008, a triumph for the underdog which he labelled 'a fairytale' and which provided an indication of a talent that would result in four world titles.

In 2020, at an eerily empty Monza due to the pandemic, Pierre Gasly delivered a shock win for AlphaTauri.

'I have spent a lot of my life in Italy and now I live in Milan, so of course, it's a very special place for me,' says Gasly. 'It's a semi home race for me since I stay at home for the race weekend, so that is very nice.'

After Gasly's surprise victory he sat alone on the top step of the podium after the ceremony, covered in ticker tape and champagne, in disbelief. Grand prix winners make up an exclusive club: in the 75-year history of Formula 1, there have been just over 100.

'I had a lot of thoughts and a lot of emotions,' he recalls. 'And I was about to walk out of the podium and I said, *OK, I've waited for all my life for this specific moment.* I mean, there's only one first win, that specific feeling of the first time you're achieving it. And I was like, *OK, just take every second, this is your moment.* If I want to stay there 10 minutes, no one will tell me off for doing that. I just sat down. I had so many emotions. I just tried to take everything in. And really take a moment for myself, because you're working so hard for such a specific moment that it's important to enjoy and take the time to do it. We're always thinking what's next, what's coming next, all the time, but sometimes we forget even to enjoy the moment that we're living in.'

Gasly was unfortunate enough to claim his victory in a year in which fans were locked out of Monza because of the pandemic. Monza's podium is the most frenzied and celebratory of the entire season. It has always been the scene of passionate support and post-racetrack invasions, but that was amplified at the turn of the century when the circular podium was relocated to hang across the pit straight. It means the area beneath the podium is crammed with enthusiastic fans, who clamber through and under the fences straight after the chequered flag, while those lucky enough to stand atop the podium for the trophy presentation are greeted with a red sea of flag-waving and flare-carrying spectators which stretches along the length of the pit straight. Those not representing Ferrari are occasionally, and unfairly, booed, though those who have delivered a home podium are universally lauded, the chants continuing through the playing of rival national anthems. And if the winner is a Ferrari driver, well, the reception is akin to Italy winning a World Cup final as the upbeat Italian national anthem blares out and is sung at full gusto. A Ferrari driver competing at Monza is an icon, but a Ferrari driver *winning* at Monza – that's legend status secured there and then.

The chaos and passion of a Monza podium epitomises the grand prix.

Monza is not the best organised event, with hand-waving and shouting where it is not required, no one to be seen when it is, and access tricky on account of its setting within a park. But to an extent, Monza gets a free pass for reasons of sentimentality, its location (not to mention the generous servings of pizza and gelato) and the tacit understanding that it is irreplaceable.

'The Monza commentary box was by far my favourite because it was terrible!' says commentator Alex Jacques. 'It's been condemned now, it's no longer safe, you had wires out of the wall, incomplete floorboards, incomplete walls, you had windows with bits missing, but it was my favourite, a lovely art deco grandstand there since the days of Alberto Ascari in the '50s. Very rarely in motor racing do you end up with something that's in every camera shot, as there's a lot of modernisation. Silverstone doesn't have a building that's been there from 1950 onwards, but Monza does – and because of that history and the wonderful turret on top, if Ferrari win, the view from there, with all the *Tifosi* on the straight, it's got to be one of the very best views in Formula 1.'

However, there have been subtle indications from Formula 1 that Monza cannot rest on its laurels and rely solely on its history.

'OK, this track is special and we know it,' explains Valsecchi. 'But to maintain and to keep Formula 1 and doing the events, we need to invest a bit of money. You know Silverstone? It's amazing, this place; the old pit lane was all crap. And now it's becoming perfect, so we need to do something like this. I mean, OK, you have the circuit that is iconic, but sometimes it's not enough. You know, nowadays it's not enough. But I hope it's going to stay on the calendar forever.'

Fortunately, there was investment from the state of Lombardy and from the Automobile Club d'Italia and in early 2024 a three-year project got underway to upgrade some of Monza's facilities, with revisions to grandstands, access tunnels and hospitality areas.

'We want to combine the extraordinary history of this facility with technological research and the most cutting-edge architectural solutions, naturally in the utmost respect for the iconic place in which we are,' said Angelo Sticchi Damiani, President of the Automobile Club d'Italia, in a press release. 'We are inside the second largest fenced park in Europe, a restricted park, a treasure to be safeguarded for future generations.'

Bidding farewell to Europe for the season means saying goodbye to the motorhomes – and their associated crews – for the year. After an intense usage across the continent between May and September, they disappear for seven months of the year, but there is still work to be done thereafter.

'For the nine races we use it for, we obviously log any damage, anything that we need to repair or replace, or if we see a hydraulic leak or a water leak,' says McLaren's Trackside Operations Director Mark Norris. 'It's only doing nine races but it's constantly travelling during that time, it's taken up and down continuously, and you get wear and tear on that.'

McLaren's Team Hub returns to motorhome manufacturer Bischoff + Scheck, in Germany, where it undergoes an extensive winter maintenance programme. The individual pods are worked on separately, laid out, and towards the end of the process the entire motorhome is rebuilt inside the company's warehouse. That enables a full-scale assessment of the facility while also allowing tests such as dousing the roof in water to check for any leaks.

'They are hammered quite hard in those nine races,' Norris says. 'You've got lots of footfall going in and out of there. You'll inevitably get something that fails, maybe something electrical, but we go through testing, service the air conditioning, go through hydraulic testing.'

It would feel wrong for August to slip away into September, with Europe basking in the last throes of summer and for Formula 1 not to be at Monza. Including Imola, the one-off Pescara Grand Prix in 1957 and pandemic stand-in Tuscan Grand Prix (at Mugello in 2020), Italy is the only country to have held over 100 grands prix and Monza is still the Don. It used to be a race held in the closing stages of the season, but amid the expansion of the calendar there now remains a third of the year left to run, all involving lengthy flights across the world.

19

AZERBAIJAN GRAND PRIX: BAKU CITY CIRCUIT, BAKU

> *'We've got to be on top of our game in all areas, whether it's nutrition, recovery, training, mental preparation, it's just very draining throughout the whole year.'*
>
> PIERRE GASLY

One of Formula 1's more obscure locations takes the championship to a country known as the Land of Fire, in a windy city which lies below sea level and where the circuit has a castle wall as its apex.

Azerbaijan's capital, Baku, is one of Formula 1's more unusual venues and locations, the championship plunging into the very heart of the city – smoothed by extensive government backing courtesy of the oil wealth that underpins its economy – and incorporating some of the primary landmarks. It is the lowest-lying race, around 30m (98ft) below sea level, adjacent to the inland Caspian Sea, and is colloquially labelled the City of Winds for its frequently gusty conditions. Preserved ruins, some almost a thousand years old – such as the Maiden Tower, a UNESCO World Heritage Site – sit alongside imposing Victorian and Soviet-era architecture, Parisian-inspired boulevards and futuristic developments, most notably the sweeps of the Heydar Aliyev Centre and the glass-fronted curved Flame Towers, complete with LEDs to create the illusion of flickering fire. It is a city at a crossroads of the world that wrestles with various identities.

Parts of Baku resemble Dubai, a walk around the corner resembles Russia, and a walk down further alleys takes you to crumbling and ramshackle community housing, some of which has been razed by the regime to redevelop the city's image. There's also a dollop of London thrown in, with an array of black cabs, albeit painted purple.

Formula 1 arrived in Azerbaijan in 2016, initially taking on the moniker of European Grand Prix. The race was won by that year's eventual champion, Nico Rosberg. A year later, the event reverted to its country's title, with an action-packed grand prix won by Daniel Ricciardo, who surged past rivals to triumph despite starting a lowly 10th on the grid. Formula 1 represented Azerbaijan's biggest annual sporting event during an era in which it sought to use sport and entertainment to align itself with Europe, holding the European Games in 2015, the UEFA Europa League final in 2019 and select games of UEFA Euro 2020. Its placement in the schedule has migrated even in its short history – held in June, then April, June, April, before finally shifting in 2024 to mid-September.

Formula 1's paddock sets up shop in front of the imposing gothic Government House on Freedom Square, in the shadow of the modern LED-fronted JW Marriott Absheron and Hilton hotels – which are at least trustworthy, an important consideration given the events of one year, when the reception at one digs simply disappeared overnight to leave an empty corridor. This is a tightly knit paddock, and the only VIPs and celebrities in attendance are those remunerated to perform at the concerts; both Dua Lipa and Christina Aguilera have been incentivised to go to Baku.

This is also a grand prix that can be tricky to reach owing to a lack of direct flights, despite Baku's close proximity to Europe, meaning convoluted journeys at erratic hours. Everyone has a tale of a random airport at which they have connected to reach Baku: Doha, Istanbul, a pre-war Moscow, a pre-war Kyiv, Budapest, Frankfurt. It is almost a game to see who had the worst and most unusual connection with an unexpected airline. It is one of the few grands prix where most teams will charter a plane for convenience.

The circuit incorporates Baku's different imagery.

The first section of the lap incorporates long straights, heavy braking points and 90-degree corners through tree-lined residential avenues, before it plunges into older streets, circumnavigating the fortressed Old City. That includes the narrowest portion of a Formula 1 track all year: the tricky left-hander at Turn 8 is just 7.6m (25ft) wide between the barriers and the castle walls, and its cobbles must be covered ahead of every grand prix. A fast and flowing section, including multiple blind corners, leads drivers back towards the Caspian Sea and on to Neftchilar Avenue, repurposing one of Baku's wide arterial roads into a full-throttle section of racetrack for the weekend. That incorporates several high-speed kinks, lasting almost 2km (1.2 miles), which has led to some multi-car slipstream battles across the years, as well as some dramatic accidents. Baku's first six grands prix delivered six different winners, before Sergio Pérez finally doubled up in 2023, demonstrating the whacky and wild races possible around the unusual circuit.

Baku is now Formula 1's sole outpost in a region where the championship has not fully found a foothold, meaning it can feel slightly isolated.

Turkey was part of the calendar from 2005 through 2011, at a superb permanent circuit on the outskirts of Istanbul, but the event was unable to attract a sufficient crowd nor match Formula 1's financial expectations over the long term. It returned as a surprise stand-in during the pandemic across 2020/21, delivering a pair of enthralling races in slippery low-grip conditions, and was the location for Lewis Hamilton to clinch his seventh world crown, but fell by the wayside once more when normality resumed.

Russia had several false starts and failures to attempt to secure a Formula 1 round, and finally in 2010 a seven-year deal was announced to hold a grand prix in the country at Sochi Autodrom. The new semi-permanent circuit was designed around the stadia that had hosted Russia's 2014 Winter Olympics in the Black Sea resort of Adler, a short hop from the Georgian border. It had little in the way of redeeming features, the race was usually unspectacular, and the venue was fairly disliked on account of its soulless nature and dour location. Sochi, or more specifically the district of Adler, had the vibe of an out-of-season holiday camp and neither the hotels nor restaurants offered efficient customer service. There was a theme park with hardly anyone there and newly built road networks for barely any traffic, which were plonked alongside existing residential facilities.

Formula 1 made few in-roads in Russia, but it was a useful ally in terms of finance and Russia's President Vladimir Putin occasionally attended to preen at the event. Sochi was set to drop from the schedule after 2022, following the decision to move Russia's grand prix to the permanent Igora Drive facility near Saint Petersburg, which more closely fitted Formula 1's ambitions of holding events in the periphery of renowned cities. This raised hopes that Russia's race could extricate itself from the malaise of apathy and be injected with a new lease of life, amid plans for a midsummer race to take advantage of lengthy days in such a northerly location. But Formula 1 never made it to Igora Drive. Russia's invasion of Ukraine in February 2022 prompted Formula 1 to immediately axe that year's event and it swiftly terminated its contract with the promoter of Russia's grand prix. Few tears were shed.

By the time Formula 1 leaves Europe and begins to venture further afield, jet lag re-emerges as a factor for the paddock. The season is bookended by a sequence of 'flyaway' grands prix that involves long flights, connections and huge time shifts, often in quick succession, the most brutal stretch of which is at the end of a campaign. This is when a certain seat in an A380 or 777 begins to feel as comfortingly familiar as a hotel room, doubling up as an office and a bed; when airport terminals begin to blend into one; and when the hour of the day becomes something of a construct. The routine of airport/plane/hotel/circuit is a cycle that can become normalised.

Over the course of a season, teams and drivers will take in excess of 50 flights solely for the purpose of getting to grand prix – before other marketing days and commitments are factored in – and spend the equivalent of over a dozen days on aircraft.

'I feel like I've been jet-lagged for a month-and-a-half,' quipped Kevin Magnussen towards the end 2023.

To extract the maximum level of performance from top-level athletes, combatting jet lag is crucial.

'Internally we all have a circadian rhythm, which is a biological clock that our body adheres to, and this affects many different systems. Metabolic rate, various hormones, core body temperature, and how that is perceived by us on a human level is our level of "awakeness" or sleepiness,' says Rupert Manwaring, a Formula 1 Performance Coach. 'These guys are operating at a very high level, physically and mentally, the cognitive demands are high, the G-forces are high, in very environmentally demanding places, so what we need to do is make sure the driver is in the best place physically, as part of their body clock rhythm, for when they jump in the car.'

The simple rule for any traveller is to allow for being able to shift one hour per day, but there is no definitive solution, and it can depend on the person involved and how their body reacts to such large shifts.

'If it is a nine-hour time difference, we'll try and arrive that number of days in advance, but that can be a challenge over the course of a season, as home time is important,' says Manwaring. 'From a performance perspective it'll be great to nail everything by the book, but we are dealing with humans, not robots.'

Drivers, most of whom are based in Western Europe, will try to get into a routine across the course of a season.

'For the European races, usually it's Thursday morning to Sunday evening, so it's a quick trip,' explains Valtteri Bottas. 'For the "flyaways" it depends on the time difference. For Australia I want to be there the weekend before, the Saturday before, so you get almost a full week before you start driving and need to be on it. The races in the US, quite often Monday or latest Tuesday, the same with Asia, I wouldn't go any later than Tuesday, just to get into the rhythm. You always learn something; you learn if you've gone too late to a race with a time difference, if you made a mistake because you didn't sleep well.'

Manwaring explains that 'the most important thing is to nail that first night's sleep, get the journey right, and get the preparation right. The jet lag symptoms last between three to five days, but negative performance effects can be between seven and nine days, and we may not realise it. The jet lag symptoms are quite strong

– disorientation, energy drop – those are the symptoms we are all familiar with, but performance symptoms you might only notice when doing a high level of performance, as that's not something we notice on a daily basis, so we have to factor that in as well.'

For long-distance journeys jet lag is combined with travel fatigue, an aspect that can actually be helpful, particularly when trying to adjust to new surroundings.

'In some cases it is used as a tactic, to deny yourself sleep, so it makes it easier to fall asleep,' says Manwaring. 'It can help in those initial couple of days. Where it is an issue, and if you've left your travel too late, is when that travel fatigue is an issue on the days you're looking to perform, and obviously that is something that can be an issue!'

Ensuring the right environment for a good night's sleep is also crucial. Drivers don't want to be kept awake by loud external noises, parties that may be ongoing in a city, or risk being in uncomfortable surroundings.

'The biggest thing for us: *Is the bed decent straight away?* says Jon Malvern, Founder of Pioneered Athlete Performance. 'It's important to have a cold room. During the night your body temperature starts to reduce. When it starts to reduce that's when you can get into good REM [Rapid Eye Movement] and deep sleep. That's what your body associates with *OK, I'm in a safe place, I can relax now, I'll turn everything off.* And generally cooling helps reinforce that effect. If you're somewhere hot and your core body temperature is warmer, it creates a challenge to try and get into deep sleep. So you just have a light sleep and you end up waking up a lot, which is obviously not good for recovery. Sometimes we take our own beds or sleep kits. You actually take something like a mattress topper and your pillow that could go on top of any surface and create something similar.'

Choosing the right location in which to stay for a weekend is also vital.

'At all the European races I stay in a motorhome, which is parked close to the track, so it's nice and easy,' says Bottas. 'At all the rest I have my favourite hotels; some in the last few years I've had to change because of the increased traffic that F1 is experiencing thanks to the popularity; it can be trickier to get in and out! I try and prioritise being closer to the track than in the middle of the action.'

There are other mechanisms that can be utilised to adapt to the new time zone as swiftly as possible.

'It's a combination of what in theory provides the best potential to perform well, not eating anything that could be performance-limiting, and also what makes you feel good,' explains Malvern.

Most drivers tend to keep to the same meal routine at grand prix weekends – knowing that it works and eradicating the risk of an unexpected differentiator. Drivers typically consume between 3500 and 4000 calories per day during a grand prix weekend in order to have sufficient energy to perform at their maximum.

'Driving a car is very mentally taxing, your brain is hungry for carbohydrates, and very fussy on what it likes to use for fuel. You've got to make a lot of decisions, make them quickly. You've got to learn a lot over a weekend. You also want things that aren't likely to take a long time to ingest or make you feel lethargic, and minimal risk in terms of what can go with it in terms of like food poisoning.'

The body also must be able to tolerate taking on enough before heading out for a two-hour race, in which fluid intake is limited to whatever can fit in a small pouch, which is sometimes not installed anyway, to save weight.

The expanding nature of Formula 1's calendar, and the relentless travel, means there is a greater susceptibility to being waylaid by bugs and viruses, and everyone needs to keep on top of their health.

'I think recovery is more important than ever now in Formula 1,' says Pierre Gasly. 'We've got to be on top of our game in all areas, whether it's nutrition, recovery, training, mental preparation. It's just very draining throughout the whole year. Where before we could get away with a few things with 17 races, now when you come into Formula 1 you have to give up a lot of your personal life and accept it. It's just being aware of that and finding the right balance to get the maximum out of yourself.'

The expansion means that the 2024 season has 50 per cent more grands prix than 25 years previously. Azerbaijan now acts as a transition between Formula 1's European heartlands and the true start of the 'flyaway' rounds, with some fan favourites and fresher destinations still looming on the schedule.

Baku has many different faces, while its circuit has the narrowest section of track on the calendar

20

SINGAPORE GRAND PRIX: MARINA BAY STREET CIRCUIT, SINGAPORE

> *'It's just quite uncomfortable driving, like you always feel just very warm and you have to get used to just sweating and it can't really go anywhere, it's just in your suit, and so you have to just get comfortable with that.'*
>
> MAX VERSTAPPEN

Sitting in a stiflingly confined hot bath that's gradually filling with your own sweat, while being shaken across the bumps of the city streets for almost two hours – that's the Singapore Grand Prix.

The long-haul legs recommence for Formula 1 in the Lion City, a vital trading post perched on the edge of the Malay peninsula, and a snapshot into the vast and vibrant cultures of Asia. Formula 1 turns off the lights and sparkles in the city of Singapore.

Singapore had a brief flirtation with motor sport in the 1960s and 1970s, when the fast sweeps of the leafy Thomson Road area held Formula 2-level events, but the gathering was eventually discontinued over safety concerns: seven racers died across a decade.

Formula 1 signed a long-term contract with Singapore, supported by the city state's tourist board, to hold a round from 2008. A course was mapped out around the Marina Bay district of the city and this broke new ground for Formula 1 by becoming the first night-based event in the championship's history. Huge scaffolding was erected on one side of the circuit and 1500 light projectors installed, placed a few metres apart and providing around 3000 lux, significantly brighter than ordinary street lights. This creates near-daylight conditions while the lights are angled to prevent potentially disruptive glare in wet weather. The night-based nature of the event enables Formula 1 to broadcast in a time zone that is more convenient for European audiences and, while other events have since switched to taking place after sunset or at twilight, Singapore remains Formula 1's original night race. It was absent for two years in the pandemic, sparking fears over its longevity and rumours that the state no longer regarded Formula 1 as essential, given that the sport had raised its status on the global stage, but those aspersions were swiftly knocked on the head when a long-term deal until 2028 was agreed.

Formula 1 initially had a presence in Southeast Asia from 1999, when Malaysia joined with a new circuit called Sepang, adjacent to Kuala Lumpur's international airport. Malaysia's state-owned oil and gas company, Petronas, has had a lengthy presence in Formula 1, as a partner to Sauber, and since 2010 as the title sponsor to Mercedes. Formula 1 raced in Malaysia for almost two decades, but the country opted to discontinue its event in 2017, in part because of the emergence of Singapore's more attractive and vibrant race, and amid declining ticket sales at Sepang.

Vietnam was due to join Formula 1's schedule in 2020, as the first grand prix inaugurated by Liberty Media, and a swish facility incorporating a hybrid of public and private roads was constructed on the outskirts of capital Hanoi in under a year. The championship was due to venture into Vietnam on the first weekend of April 2020 and the circuit made it as far as being included in the official video game, but the onset of the pandemic meant the event was cancelled just three weeks out. Before the pandemic passed, Nguyen Duc Chung, Hanoi's mayor and a huge influence in the event's foundation, had been arrested and imprisoned on corruption charges unrelated to the grand prix. There was consequently little appetite in Vietnam to revive the event and thus the facility – including a permanent pit lane and paddock building – sat fully built but unused and dormant. There has never been any indication that Vietnam's race will be resurrected.

Singapore's circuit is among Formula 1's slowest, and bumpiest, with 19 corners spread out across its 4.9-km (3-mile) layout. There used to be more turns, but an awkward high-kerbed chicane that launched cars skywards if drivers erred, colloquially dubbed the Singapore Sling after the famous cocktail created in the Long Bar of the nearby Raffles Hotel, was removed after five years. Four corners were then axed in 2023 when a section of track which ran in front of – and beneath – an enormous grandstand was bypassed due to renovation works by the city. The circuit carves its way through the city, past or beneath some of Singapore's major landmarks or highways, and Formula 1's event is undoubtedly the greatest tourism advert for a city that acts as a gateway to the region.

Access is convenient: the acclaimed Changi Airport is a short taxi ride away, while Singapore has a well-connected metro network, the MRT, and several stops and lines cover the major access points to the track. The MRT also came in handy one year when Esteban Ocon used it to return to the paddock after retiring from the race at the far end of the track. There have also occasionally been track invaders of the reptilian type. Practice sessions, which take place in the daytime, have been disrupted by the presence of monitor lizards – common in the city's green spaces – wandering on to the circuit.

Singapore is the event that drivers have in the back of their minds when undergoing gruelling training regimes throughout the season. If you can cope with Singapore, you can probably cope with anywhere. It isn't necessarily the heat – temperatures usually peak in the low 30s (Celsius; 80s Fahrenheit) – but rather the humidity, which is normally at least 80 per cent. Anyone unaccustomed to the city will find that even just an amble through Singapore's suffocating streets will leave you soaked in sweat, and a pedestrian should seek out the shortcuts through air-conditioned shopping malls and arcades. Drivers can lose as much as 4kg (almost 9lb) during the course of a 62-lap race.

'You lose a lot of liquid, you sweat a lot and it's very demanding of the driver,' says Kevin Magnussen. 'In all the training I do, the main motivation to train is Singapore. That's how I feel anyway. The most exhausted I've ever been is in Singapore. There can be good and bad years with the humidity, but generally it's the most demanding race.'

In the build-up to Singapore drivers will try and train in the heat of the day in Europe, before flying out as early as possible to acclimatise, or train in heated rooms indoors – such as saunas – while wearing multiple layers of clothing, including the likes of jumpers, fleeces and leggings. That can get some strange looks when out in public, but it is a necessary aspect of preparation.

'Singapore's really tough for humidity, and humidity is actually a really interesting challenge,' says Jon Malvern, Founder of Pioneered Athlete Performance. 'The best way to balance that is just to be supremely fit, exercising regularly because you get used to dealing with heat, and then you can layer up things like working out in heat, in a humidity chamber, or working out in a sauna.'

Cockpit temperatures can exceed 60°Celsius (140°Fahrenheit) in Singapore and the ambient heat is energy-sapping and oppressive, accentuated by the absence of any breeze, given the circuit's location in the concrete-laden streets and between high-rise buildings. Opening the visor for a little bit of air is unlikely to feel anything other than unpleasant.

Drivers are surrounded by tight confines and equipment that can also overheat in the conditions, while Singapore's circuit layout means there is little margin for error and little in the way of respite. Singapore's slow average speed also means it is often a long race – it has hit the two-hour time limit on multiple occasions – and that enhances the challenge. The deployment of the safety car – which has a 100 per cent presence record in Singapore – can offer a slight breather.

But the unique aspect of motor racing is that sweating – seen as a benefit in most sports – is a challenge that needs to be overcome.

'In any other sport, you're out on the field or you're cycling or you're running, you get hot, and in a bid to control your core body temperature and to cool down you send blood to the surface of your skin,' Malvern says. 'So you go red. You start producing sweat to whip heat away from the skin, which cools the blood and keeps your core body temperature in check, and it makes you feel okay. But when you're racing, you've got fireproof overalls, Nomex and you're enclosed. So sweating is actually, in theory, a negative.'

Controlling that inevitable process is a challenge.

'It means you just lose fluid mass, which means your blood gets thicker, which makes it harder for you to pump the glycogen you want to get around your body and the oxygen around your body to fuel what you're doing,' Malvern says. 'You can do things to try and combat that with sodium intake, for example, so you retain more water generally and you've got more capacity to lose water before it starts becoming detrimental. But that's quite hard on the gut, very tough on the gut. You have to train yourself to be able to do that.'

It is not a pleasant experience for drivers.

'It's just quite uncomfortable driving, like you always feel just very warm and you have to get used to just sweating and it can't really go anywhere, it's just in your suit, and so you have to just get comfortable with that,' says Max Verstappen. 'Yeah, it's a tough track. You need to keep focused throughout the whole race. It's easy to hit a wall here and there. But that's always tough on the street circuits, but then of course with increased heat, just trying to stay focused for almost two hours.'

Pre-Singapore is all about getting as hot as possible, preparing and acclimatising the body to the oppressive conditions, but once the weekend starts the focus shifts, the aim now is to keep the body as cool as possible.

Teams will have ice baths ready for drivers between sessions – sometimes as rudimentary as a wheelie bin or a big bucket full of ice cubes – while cooling vests, which have pouches filled with ice, are commonplace because this also lowers body temperature while assisting recovery. Drivers will also have multiple showers a day and a cold shower is favoured shortly before getting in the car for each session. Sugary soft drinks are also taken onboard to help restore lost nutrients: one driver favours Coca-Cola, often downing several cans after the race. Drivers can install a drinks 'bottle' – effectively just a little pouch that connects to their helmet – but some opt out to save weight and those that do carry them will find any refreshing drink tastes pretty unpleasant after just a few minutes.

'I think it's more a tea that we have to drink during the race to rehydrate because it doesn't taste good at all,' says Carlos Sainz. 'It's not what you want to have when you're struggling in the heat.'

It is an event at which the drivers' trainers and physios earn their crust.

'I'm forcing lots of water to keep him hydrated, over-hydrated,' says Nikolaj Madsen, Kevin Magnussen's physiotherapist. 'After the race we try to put some sugar inside him because the brain is overheated, and the brain works with sugar. As fast as I can, I try to get some sugar inside him, and slowly some water. After a long race, you don't tend to be hungry because the adrenaline is still in your body and sometimes it's hard to sleep after a race.'

While drivers need to be on top of their game, so too do team members, who have to work in extremely challenging conditions in the garages and the pit lane. They have to carry out their roles, just as at any grand prix, which means wearing fireproof suits and helmets during pit stops.

'Avoiding heat illness is our priority,' says Faith Atack-Martin, Haas F1 Team Performance Coach. 'This relies on maintaining equilibrium between salt and water balance in the body. In hot and humid conditions, the surrounding air is higher in water molecules, making it harder to regulate body temperature because heat cannot evaporate as efficiently. The body responds to this by increasing the rate it sweats, resulting in increased water and salt loss. . . .

'Cooling strategies include access to iced towels, sweat bands and the use of an air-conditioned breakout room in the garage where team members can go to take a break. We also use menthol mouthwash. Although using this does not affect core temperature, the perception of heat can be manipulated. Similarly, using cold sprays gives a temporary feeling of coolness that can help in addition to more robust methods.'

Small but crucial details can be influential. One team manager – new to their position – arrived in the pit garage in Singapore and queried why ordinary office fans were being used. A couple of people were duly dispatched to find industrial-strength air conditioning units, which the majority of rival teams already had and which gave the mechanics significantly more pleasant conditions in which to work. However, the trade-off of having powerful air conditioning is the risk of replicating cold or flu-like symptoms, meaning there is no perfect solution – and this is a battle to be fought at increasing numbers of races as the calendar expands and more events in hotter climates join the fray.

One factor with Singapore's grand prix – which is no longer a unique aberration amid the arrival of Qatar, Las Vegas and other evening-based events – is the timetable. The race begins at 20:00, meaning personnel effectively remain on European time, their waking hours roughly 13:00 through 06:00. However, this is not a strict rule, given that some prefer to maximise daylight hours and some team departments – such as marketing – need to align with partners that are on normal Singapore time.

'It is very hard to deny the body its environmental signals, which will pull the body towards the time zone,' says Rupert Manwaring, Formula 1 Performance Coach. 'Light is a big factor: artificial light is still effective in trying to keep the body in a specific time zone, as light is associated with wakefulness and our bodies interpret it as such, and we can use that to manipulate us towards a certain time zone. But it is difficult in Singapore because ultimately the race is late, then you have the debrief, other commitments, and it is dark, it's hard to deny the environmental signals that the body is being told, *It's time to sleep*.

'There's things like we wake up at 1 p.m., 2 p.m., and outside there will be soundchecks for the events going on that weekend. At that point the hotels and the local community are not necessarily factoring in we're trying to sleep, or hotel cleaning services knock on the door at 10 a.m., silly things that can play a part. There's things you're constantly battling with the environment. Even for an evening person, that's still quite late, an 8 p.m. start, so you have to make a big effort to stay on European time. Come Monday, you are still fighting Mother Nature to stay on that time zone – it is a difficult thing to do.'

It often means personnel congregating across Singapore's hawker centres in the early hours of the morning for dinner, to take advantage of the 24/7 street food stalls, most restaurants having long since closed up for the night. Fortunately a cluster of international hotels frequented by teams near the track have since learned to be accommodating and anticipate gaggles of mechanics arriving at atypical hours in pursuit of food.

'The biggest challenge is going back to the hotel when daylight breaks,' says Aston Martin Performance Director Tom McCullough.

'You don't want to be taking in too much natural daylight before you go to sleep, as three sessions take place under darkness. Sometimes, after qualifying, it's easy to think about lots of extra what-if scenarios, get back too late and walk back in the daylight. That plays havoc with your sleep pattern.'

Despite the heat, the humidity and the awkward timetable, Singapore has established itself as a popular event on the calendar, with the multicultural city one that truly embraces the grand prix. Something of a community spirit is fostered when everyone congregates at the same hawker markets at 3 a.m., reacquainting themselves with the locals who run the stalls. Singapore's placement in mid-late September means it is regularly the first post-summer, non-European event and marks the start of the 'flyaway' season, beginning the gradual slide towards the end of the year.

The illuminated Marina Bay circuit slices through the heart of Singapore

21

UNITED STATES GRAND PRIX: CIRCUIT OF THE AMERICAS, AUSTIN

> *'Often when they build new circuits, they're not really like the classics. But this is a classic from the get-go. And it's one that enables us to have great races.'*
>
> LEWIS HAMILTON

Austin ticks several boxes for Formula 1. It is a spectacular circuit in a vibrant yet compact city that is renowned for its bars, food and live music scene, and where attendance has reached almost half a million people in a country where the championship is eager to expand its foothold.

That wasn't always the case – even in Austin.

Formula 1 has a long and chequered history when it comes to the United States Grand Prix.

Miami joined Formula 1's calendar in 2022 and Las Vegas arrived on the scene in 2023, but since 2012 the country's national grand prix has taken place at the Circuit of the Americas, to the southeast of Texas' distinct capital, Austin.

Florida's Sebring, renowned as a sports car mecca, hosted the inaugural world championship United States Grand Prix in 1959, a race won by McLaren founder Bruce McLaren, for his first victory in the championship. The US Grand Prix switched to California's Riverside in 1960, where Sir Stirling Moss triumphed. Then

Formula 1 set up home at the permanent Watkins Glen facility in upstate New York in 1961.

The hilly Watkins Glen remained on the schedule for two decades, but as the 1970s progressed Formula 1 began to outgrow the facility, despite an extension to its layout, while a lack of money meant it dropped off the radar after 1980's event.

The US GP moniker was absent for almost 10 years through the 1980s, but there were street events in Long Beach (eight times, hosting a race that became a CART/IndyCar staple once Formula 1 left), in Dallas (seven times) and in Detroit (just once), as well as an unsuccessful two-year attempt to create a long-term future in Las Vegas. The US GP name returned in 1989, with a street event in Phoenix, Arizona, won by Alain Prost, but the uninspiring circuit was never truly popular – and attendance was low (estimated at no more than 40,000) despite opening the seasons in 1990 and 1991, when Ayrton Senna began his eventually successful campaign. Phoenix never rose and that left Formula 1's US dream in ashes.

At the turn of the new millennium Formula 1 returned to the US and this time rocked up at the Indianapolis Motor Speedway. One of the largest crowds in Formula 1 history, believed to be around 250,000 spectators, attended Indianapolis' 2000 debut, yet the sport still never truly took off.

While Formula 1 utilised a stretch of Indianapolis' famous oval it remained in the shadow of the Indianapolis 500, and the road course section of the circuit was flat and uninteresting. But the biggest misstep came in 2005 when Michelin, tyre supplier to seven of the ten teams, could not guarantee the safety of their product through the banked oval after a spate of inconclusive failures through practice. Lengthy discussions took place and remedies were proposed, including the installation of a temporary chicane on the oval, but no satisfactory resolution was achieved. The outcome was 14 of the 20 competitors pulling into the pits at the end of the formation lap, leaving only the Bridgestone-supplied cars (just six) on the grid to contest the race. Unsurprisingly that was not exactly a public relations success for Formula 1. Only two more editions were held at Indianapolis, Lewis Hamilton delivering victory across the yard of bricks for the last time in 2007.

In 2010 the US Grand Prix was revived and for the first time a circuit would be specially constructed for Formula 1, the deal announced for 2012, in Austin.

'I got involved because I owned some property to build houses on out here,' says Bobby Epstein, Chief Executive of Circuit of the Americas (COTA). 'A couple of guys came and said, *Maybe you can sell your land or get rid of your land, if you help them do a business plan to bring Formula 1 to Austin.* And I had seen Formula 1 race earlier, and my family is from Indianapolis, so I loosely followed racing growing up and was familiar with Formula 1. I thought that'd be a really cool project and when you start to look at what it can do for the economy, where you live, especially coming off the great recession of 2008 and trying to create jobs, it seemed like a great opportunity in that we could do some good things for where we live – and we might be able to make some money out of it!'

There were some early teething problems, and contractual disputes, prompting Epstein to take on the lion's share of the development work to avoid the project falling through in 2011. Formula 1 arrived for the first time in November 2012 and the circuit was immediately popular among the drivers for its range of corners and the ability to battle wheel to wheel, while the allure of downtown Austin was another boon. Even so, there were a few difficult years, the nadir coming in 2015. While the race was a thriller, with Lewis Hamilton clinching his third title, the weekend-long wet weather caused havoc. There was flooding, damaging some infrastructure and turning the venue into a mud bath. It was a confluence of setbacks arriving at once and financially detrimental for a venue that was still in its infancy and still trying to convince locals – especially local taxpayers – of the benefits of the event. Fortunately COTA steadied the ship with a spate of smoother events and as the 2010s trickled towards the 2020s there was an uptick.

This was caused by a combination of factors. COTA's 2021 race was the first outside Europe and the Gulf since the pandemic, coming at the height of the Max Verstappen–Lewis Hamilton rivalry, and shortly after the Netflix documentary series *Drive to Survive* had taken off in the country. The docuseries was launched by Liberty Media to cover the 2018 season and initially only eight of the 10 teams approved their involvement. The first season debuted in early 2019 and the non-participants, Mercedes and Ferrari, swiftly cottoned on to the reach of *Drive to Survive*, a realisation that accelerated when people were searching for programmes to watch during the pandemic – by which time there were two seasons to watch.

Liberty Media's greater social media presence was also paying dividends, enticing a younger and broader fan base, particularly in the United States.

Attendance had gradually grown, reaching 268,000 in 2019, but just two years later that figure had surged to 400,000 and it reached new heights in 2022, when 440,000 spectators attended the weekend's action.

'COTA put F1 first,' Epstein says. 'It wasn't a by-product of another sporting event, whether it's racing or other sport, it was simply built for Formula 1 and I don't think there had been another [US] venue that was built specifically for them. That was important to Bernie Ecclestone at the time. That was . . . one thing they thought was [important] – besides the fact that his main concern was that the cheque would clear! – but second to that was the fact that you're actually doing something that was F1-specific. You could point out some of the cases where other races either today or in the past are . . . secondary to another business that company has, or adapted to be a Formula 1 circuit. Even the case of Indianapolis, where their identity was with IndyCar and the Indy 500, and they adapted their track to create something that would accommodate F1 rather than what we've got, which is an F1 track that accommodates other things.'

COTA accommodating other things is most prominent when it comes to music acts, with the venue one of the first to tap into the value of in-weekend concerts. The idea was not only to feature high-profile music artists, but to showcase Formula 1 to an audience that may have previously had little or no interest in motor sport. Saturday's timetable was shifted slightly later to entice the racegoers to stay for the concert, while also attracting the concertgoers to come and check out the sport.

'Certainly the music was the biggest component, and there's no question and we keep saying it forever,' says Epstein. 'I mean that changed it. It allowed someone to bring a fan of music, and soft-sell them into the sport because they have no other way to really suddenly be near the cars and feel them, and hear them and see them, and so that was helpful, and I'm sure that we help the music industry too!'

The Friday and Saturday night concerts, on a big stage erected alongside the back straight, have become a staple of US Grand Prix weekends. Elton John headlined in 2015, Taylor Swift followed in 2016, while Stevie Wonder, Justin Timberlake, Britney Spears, Bruno Mars, Queen and The Killers have all performed the main weekend concert. That also aligns with Austin's thriving live music scene, where a plethora of bars and cafés welcome blossoming artists, and on a larger scale the city hosts the South by Southwest and Austin City Limits festival, the latter of which typically takes place a week ahead of the US Grand Prix.

COTA's popularity was aided by its circuit, which has become a modern classic, inspired by sections of other venues to create a layout that is enjoyable for drivers and good for racing, complete with strikingly patriotic red-and-blue, Stars and Stripes run-off. The first turn is a hairpin, featuring a rise of 30m [98ft] in just a few hundred metres, presenting drivers with a wall of tarmac and the sight of the enormous US flag fluttering behind the fence. The majority of the opening half of the lap mimics the best of Silverstone and Suzuka, with high-speed changes of direction, allowing a Formula 1 car to perform at its jaw-dropping best. The second half of the lap includes a lengthy straight and fiddlier technical section, which allows for side-by-side battles. There's also one of the track's signature sections, the triple-apex right-hander beneath COTA's striking Observation Tower, a landmark visible for miles around.

The circuit is built on soft soil on a flood plain, so is prone to greater underground movement, which has led to a bumpy track despite COTA's relative youth. Some argue that it is unfit for modern Formula 1 machinery, suggesting rallycross cars would be more suitable, while others maintain that it provides some character in a landscape where track surfaces can be billiard-table smooth. COTA has had its own skin cream in the form of grinding, which smoothes the worst sections – though this is more for the benefit of MotoGP, for whom bumps are a significantly greater concern.

'Often when they build new circuits, they're not really like the classics,' said Hamilton in 2023. 'But this is a classic from the get-go. And it's one that enables us to have great races. And, you know, when you go up to Turn 1, you see that huge crowd, it's really quite amazing.'

That success has come despite the absence of home talent in a very patriotic country that can sometimes struggle to explore beyond its borders. America ostensibly has a team in Haas, which does lean into its US heritage with special events and a spectator area dubbed Haas Hill, but it is fundamentally a European outfit. Williams has had US ownership since 2020, when the Williams family sold up to Dorilton Capital, but it remains a quintessentially British team at heart. On the driver front COTA has had only Alexander Rossi, in 2015, and Logan Sargeant, in 2023. America has not had a bona fide front-running contender in Formula 1 since the 1980s, and legend Mario Andretti was the most recent grand prix winner back in 1978, the year he claimed his world title. Nonetheless there is widespread support for a spread of drivers, most notably Lewis Hamilton – COTA's first winner in 2012 – and Sergio Pérez, owing to the proximity of Austin to Mexico.

'When I used to come out to the States, like 2007, and for many, many years, every time we stepped for that one race, you always found yourself just repeating yourself, educating [people],' said Hamilton during a press conference in 2023. 'I couldn't fully understand when I went to NFL games, NBA games, seeing how passionate the Americans are about sport, how they hadn't yet caught the bug that many of us grew up catching when we were younger. And so it's been really, really amazing to see that really a large portion of the country is now speaking about it.'

It also helps that Austin is a vibrant, student-focused and compact city, with much of the very centre walkable on foot. Those able to stay downtown or in the vicinity – given that exorbitant prices and a lack of hotel rooms have increasingly forced some to migrate elsewhere – are able to soak up the best of the city's offerings.

'All us are able to stay downtown and experience a little bit of the culture of the city, as opposed to some circuits we stay in motor homes in a paddock and don't really see anything,' says Daniel Ricciardo. 'I think that's also a really nice thing with this race.'

Austin's eclectic South Congress Avenue is a regular haunt – with Home Slice or Terry Black's BBQ a stop-off on the way back to downtown hotels – and the lengthy road eventually crosses the Colorado River, leading up to Texas Capitol building off 11th street. It's true: everything really is bigger in Texas, particularly the portion sizes. Sixth Street is a hive of activity, often rammed until Austin's 2 a.m. cut-off point, particularly if college football is in town that weekend. Austin's placement in the calendar means there's usually Hallowe'en themes and decorations – and this also extends into the more suburban areas, where the houses and yards are adorned with illuminated carved pumpkins and assorted ghostly themes. Texas is also a place of big cars, expansive skies and spectacularly colourful sunsets.

'I just like all the space. It's a huge country,' says Kevin Magnussen. 'In Denmark, everything is very close together, so when you go to the US you feel like you're on a different planet.'

That atmosphere is enhanced by Formula 1 not stepping on many other sporting toes in Austin. Austin has the renowned college football team Texas Longhorns, who play in the enormous 100,000-seater Darrell K Royal Memorial Stadium, in the University of Texas just north of downtown, and there is also the recently formed Austin FC soccer team. But it isn't represented in the NBA, NFL, NHL or MLB, which are the big four of American sport that not only dominate the landscape, but which are among the wealthiest championships in global sport, with revenues that Formula 1 aspires to reach. Austin's race often coincides with MLB's World Series and the first week of the NBA season, both massive TV spectacles, but when it comes to Austin itself Formula 1 is the big draw.

Austin has a deal until 2026, but it is now firmly the spiritual home of Formula 1 in the United States. In a country where NASCAR, with its 36-event season across the States, has a stronghold and where IndyCar's Indianapolis 500 – held on Indianapolis' famous banked oval – is quite rightly celebrated, Formula 1 now has a big presence. Austin has established a loyal fan base, it is one of the most popular venues on the calendar, and it is the type of event that Formula 1 strived for decades to create in the country. It *feels* a deservedly unashamed celebration of everything Austin, Texas and America.

'I'm super proud and relieved,' says Epstein. 'You're striving for that success, and you want that, but when you do get it, it feels really rewarding. But I think we're also primed to do even better because we learned so much, and when you have the history, you learn what works and what people are responding to, and you can include those things and learn more and you can always get better. So it's not a matter of stopping and saying: *This is great, we've reached the point we wanted to reach.* It's: *We learn more things and more ways to get better.*'

Austin is one of the locations where the relationship between Formula 1 and the grand prix is hugely fruitful and mutually beneficial. There are some drawbacks – the access routes are smaller than some locations and downtown does have a chronic lack of hotel rooms – but it is the venue in the United States that Formula 1 was trying not just to create, but then to successfully evolve, for decades.

COTA established the US home that Formula 1 sought for decades

22

MEXICO CITY GRAND PRIX: AUTÓDROMO HERMANOS RODRÍGUEZ, MEXICO CITY

> *'The atmosphere at the circuit is incredible and I'd say the drivers' parade could be the best one of the year, especially when we enter the stadium section; the noise is unbelievable.'*
>
> DANIEL RICCIARDO

In a colourful and chaotic city Formula 1 has enjoyed a renaissance in recent years, with the Mexico City Grand Prix in its third – and most successful – stint in the schedule.

Mexico first joined Formula 1's calendar in 1963 and was installed as the season finale, but dropped off the schedule after 1970. It was brought back in 1986, at the venue now named after the Rodríguez brothers. Ricardo was embedded within Scuderia Ferrari as a teenager, finishing runner-up at the 1960 Le Mans 24 Hours, and had taken a best finish of fourth across six race starts. Regarded as a rising star, Rodríguez entered the non-championship Mexican Grand Prix in 1962, in a privateer Lotus, but suffered a fatal accident during practice. He was only 20. Pedro, the elder by two years, continued competing and became the first Mexican to win a Formula 1 race, in South Africa in 1967, adding a second in Belgium, in 1970. A prolific sportscar racer, he also won the Le Mans 24 Hours in 1968, but in 1971, at the age of 31, he died in an accident at the Norisring in Germany.

Mexico's second spell on the schedule ran until 1992, when it disappeared from the Formula 1 landscape once more, and there was little suggestion for a couple of decades that the country would ever return. Formula 1 was predominantly exploring new ground in Asia, looking eastwards, but there was then a mindset shift. The creation of Austin's grand prix, in 2012, sparked the realisation that a Formula 1 revival was also possible south of the border.

'I was part of the Austin group,' says Mexico City Grand Prix Director of Marketing Rodrigo Sanchez.

'The first event in Austin, the Mexican attendance was like 60 per cent. And it was like, man, like, *What's going on?* And obviously, I think we did a successful promotion in Mexico for the Austin event, but I think it did catch us by surprise, like just the number of Mexicans that showed up, I think that sort of woke us up. I think that sort of sparked the idea that, *OK, we need a race in Mexico for sure. I mean, there's a driver, there's an interest, there's a market for it.* And that's sort of how I think it started.'

Mexico duly sealed a return to Formula 1's schedule in 2015, when the race was won by Nico Rosberg, who described the event as 'really special', while runner-up Lewis Hamilton suggested the atmosphere was akin to a football stadium. The event's return was initially supported by the government, but after a few years funding for major international sporting events was halted. From 2021, the event tweaked its name to specifically represent Mexico City after the then mayor, Claudia Sheinbaum – who was elected as Mexico's President in 2024 – recognised the importance of the grand prix, and facilitated meetings and trusts for private investors to safeguard the future of the city's pinnacle sports event.

Mexico City swiftly established itself as a favourite of Formula 1's travelling circus, the promoter bringing the city's culture and atmosphere into the paddock, facilitated by the grand prix coinciding with Mexico City's colourful and celebratory *Día de los Muertos* (Day of the Dead) festival. The promoter was voted five times by Formula 1 as the best of the season, organising expeditions for paddock personnel to explore and understand the city and its culture. That has included trips to the Arena Mexico for the Lucha Libre tournaments, a voyage to the mazy and swampy Xochimilco canals, and an evening spent learning about and cooking the best of Mexico City's food scene. Inside the paddock there is a themed village, the style of which evolves annually to reflect different Mexican provinces, and there is usually a very popular taco stall. Piñatas, a mariachi band and decorative murals all enhance the vibrancy of the event.

'My experience in F1 of going to some of the tracks is, you know, you travel to a place and you're at the hotel and then you go to the track and then back to the hotel,' says Sanchez. 'We looked into how can we bring a little bit of the local culture, the local flavour, so at least if you didn't have any time to visit elsewhere, at least you take a little bit of Mexico. So that was a little bit of our approach.'

The difficulty of exploring beyond the track and the hotel comes in the form of Mexico City's notorious traffic. Journeys of a few kilometres have been known to take in excess of an hour, as cars, vans, trucks and buses trundle along the creaking highways and block the myriad junctions, prompting endless and futile honking – and this is accentuated in rush hour or wet weather. Open up Google Maps and most of Mexico City's roads will be a sea of red. *It can't take that long, surely* – but yes, it does. That is evidenced straight out of Benito Juárez Airport, where the pick-up area is strewn with taxis pointing in several directions, which then thrust down the ramp at full speed, only to head straight into miles of congestion. Those fortunate enough to have armed police escorts – such as the drivers – can shave significant time off their journeys, but for the rest it can mean lengthy commutes.

'I haven't convinced enough people to use the subway,' says the grand prix's Managing Director Federico González. 'There's a subway station right outside of Foro Sol, we have three train stations, but I haven't been able to explain it. It's five pesos, in 20 minutes from the city, you walk from the subway into the circuit, maybe 25 minutes. The city is complicated, yes, but probably very few circuits have the subway infrastructure that we have – but we have to convince more people to use it! People love to come by car!'

But good luck even trying to use the subway; it is overcrowded at most times of the day and at peak hours carriages come and go with no more room at the inn.

That chaos somehow enhances Mexico City's event, the throngs of fans congregating around the access roads and entry points amid a siren of horns, whistles and shouts from the plethora of police and security who try to direct a constant sea of vehicles into the correct lane.

That frenzied atmosphere in and around Mexico City's grand prix has undoubtedly been aided by the presence and popularity of Sergio Pérez, nicknamed 'Checo'.

Pérez made his Formula 1 debut in 2011 and spent several years scrapping as an occasional podium-finishing underdog, honing his craft in Formula 1's midfield. Finally, at the Sakhir Grand Prix in 2020 – held on the outer layout of the Bahrain International Circuit – Pérez stood on the top step of the podium. He had recovered from a first-lap incident, in which he plummeted to the rear of the field, and carved his way through the pack to secure a stunning triumph.

Pérez, who joined Red Bull Racing from 2021, has since overtaken Pedro Rodriguez's win tally, becoming his country's most successful driver, and has sealed his place as one of the most experienced Formula 1 racers in history, making over 250 starts.

He is a superstar in his home country, his face adorning the enormous billboards present in the city, and any advertisement of the grand prix centres on his presence.

'He's a proper rock star, for sure,' says Sanchez. 'I mean, I think he's the biggest sporting athlete in the country at the moment in terms of what he's doing internationally versus the other disciplines and all other sportsmen. I think there's always been a huge appetite for Formula 1 in the country. I think there is a culture and an interest for the sport in general, but obviously having Checo is like, you know, it's a perfect cherry on the cake.'

Pérez's every movement through the paddock in Mexico City is lauded, a throng of selfie-hunters encircling the home favourite, and when he is on-track there is an eruption of cheers – as if a match-winning goal has been scored – each time his car passes by the respective grandstands, most notably the Foro Sol stadium.

The Foro Sol section of track was only introduced for 2015's return, when it was deemed that the old Peraltada curve behind the stadium – a high-speed, long-radius, 180-degree sweep – was no longer suitable for Formula 1's advanced standards. Extending the minimal run-off was impossible due to the presence of a highway right behind the circuit. This robbed the Autódromo Hermanos Rodríguez of a spectacular section of tarmac, and the replacement ribbon of circuit was clumsy and awkward, but did offer a cauldron-like viewing arena. Foro Sol has seats for 30,000 spectators and this facilitates a frenzied aura during the podium ceremony. This takes place opposite the semicircular open-air grandstands, and the winning car is placed on to an elevated ramp and then lifted on to the podium. From the top of Foro Sol it's possible to appreciate the manner in which the circuit is hemmed into the district: the skyscrapers of Mexico City are visible in one direction, the volcanic peaks of the Iztaccihuatl national park present on the horizon, and there is a hotchpotch of residential buildings just outside the walls.

'The atmosphere at the circuit is incredible and I'd say the drivers' parade could be the best one of the year, especially when we enter the stadium section; the noise is unbelievable,' says Daniel Ricciardo.

Foro Sol, constructed in 1993, also doubles up as a major music venue and is often the Mexico City stop on the world tour of internationally acclaimed artists. The 30,000 spectators in the seats can be joined by another 30,000 at ground level and on the track – which was also used as a baseball pitch until Formula 1's return – to watch the likes of Taylor Swift, Harry Styles and the Rolling Stones.

'The Foro Sol is the world's number one gross revenue ticket venue,' explains Sanchez. 'It's the venue that sells the most tickets in the world: the Grand Prix, plus the concerts. Taylor Swift did four straight concerts; Coldplay did six or seven straight concerts there, 60,000, every night. We do concerts, we do racing, we do obviously track rentals, but we also host the EDC [Electronic Daisy

Carnival] and we do 25 major music festivals a year. There's a huge appetite for events these days.'

Mexico City has an additional challenge that is invisible to the naked eye: the altitude. It sits 2240m (7350ft) above sea level, by far the highest round in the championship, with Brazil's Interlagos the second highest at around 800m (2625ft).

This has an impact on the body. Some people are more affected than others, but the altitude and thinner air means that the body has to work harder to reach the same operating levels. The predicament is not helped by air pollution in the clogged city, which is accentuated by Mexico City's placement in a basin. Team members discover that it is commonplace to be out of breath quicker, particularly when climbing stairs, and also to be more susceptible to headaches and viruses. Teams are usually braced for more personnel than usual to be sidelined or slowed by illness, nausea or simply fatigue, particularly at a time of year when energy levels are already being sapped by the relentless schedule.

Mexico City's placement in the calendar, usually the weekend after Austin, means there is limited time to adapt to the altitude, though there are still mechanisms to adopt.

'For the first few days it affects your co-ordination, so I try and go [to Mexico City] Monday because of that,' says Valtteri Bottas. 'I've got lots of advice from my coach, Antti, who used to work with the Olympic team in Finland, and he had knowledge of altitude training. You need to stay extra hydrated, not just with water but also with electrolytes. You need more carbs in the diet, as that helps to keep the fluid in, so that helps you struggle less with altitude, and some iron supplements if you start a week before can help a bit with the effects.'

Drivers will tend to build up the training intensity through the week, to become accustomed to the altitude, which will avoid overstressing the body in an environment where the immune system is more compromised.

'The best rule of thumb is to just go out there as early as you can,' says Jon Malvern, Pioneered Athlete Performance Founder. 'You can go and do things like go and exercise in an altitude chamber with an altitude mask. You reduce the level of oxygen content and it allows you to start acclimating. So your body starts that process of, *Okay, I'm going to use more red blood cells, I'm going to get used to feeling okay with less oxygen.* When you get there, you just take it steady. You do a decent cardiovascular session, just anything to help you adapt quicker, but don't go in and push yourself into the red, because you're not going to have the effect you want. You're going to tire yourself out and it'll take you long to recover. It hits your immune system.'

The human element is one factor, but a bigger challenge for the Formula 1 teams is the impact the higher altitude has on car performance.

There is around 25 per cent less air density at Mexico City's grand prix compared to Formula 1 events that take place at sea level. This means that the cars do not produce as much downforce: though they run with the biggest Monaco-spec wings, they generate only the same level of downforce as the thin Monza-spec wings. The lack of downforce also means the cars slide more, accentuating tyre wear. The flip side is that the thinner air means there is less overall drag and top speeds can be very high in spite of the larger wings; Bottas set the record for the highest speed at an F1 event when he clocked 372.5km/h (231.5mph) in the 2016 race. The downside is the drag reduction system overtaking aid is less effective and drivers struggle even more to sit in the dirty air of a rival because of overheating. All this means that despite a very lengthy straight leading into a heavy braking zone, it is one of the hardest tracks on which to pass.

The higher altitude also means less air passes through the radiators, air intakes and ducts, so there is less cooling. Teams react by running bigger brake ducts, air intakes and looser bodywork around the rear of the cars, though this reduces aerodynamic performance and increases drag – even in Mexico City – so they have to find a suitable trade-off. The power unit's turbocharger must work harder to compensate for the lower air density and thus spins at a higher speed, while the brakes are more susceptible to overheating, which can accentuate wear or lead to them being glazed. This all raises reliability concerns and comes at a stage in the season where parts are usually becoming increasingly worn.

It can also be a tricky science for Formula 1 teams to find the right set-up: the track temperature window fluctuates more than normal, meaning a car can feel on form on one run, then disappear completely a few minutes later, leaving a driver scratching their heads over where half a second went. Keeping a Formula 1 car in its operating window is a constant challenge for drivers and that dance is accentuated in Mexico City.

'The grip here is so low because of the altitude and it's normally a track where you're never going to have a perfect car,' says Daniel Ricciardo. 'It's never going to feel grippy and awesome, so you've really just got to make do with what it is and navigate that with some patience. I do like it as a venue and as a circuit, because, a bit like Austin, it provides unique challenges, and when you get it right around there, you can make up a lot of time. The first sector is very fast, but when you get into the next part, even the first few corners and the chicane, you've got to take the kerbs. If you take too much, it can kill your lap time, and if you don't take enough, then you leave time on the table so that section is really tricky.'

During Mexico's first stint in the Formula 1 schedule in the 1960s it was installed as the final round of a much shorter season, playing host to an extraordinary title decider in 1964. Leader Jim Clark suffered an oil leak on the penultimate lap, paving the way for Graham Hill to triumph, but Ferrari responded by getting Lorenzo Bandini to cede second place to the following John Surtees, who duly became the only driver in history to win titles in Formula 1 and in motorcycle racing.

Lewis Hamilton clinched titles in Mexico City in 2017 and 2018, signalling the early end of a championship fight in those seasons, when there were several rounds still to go once the travelling circus had made the lengthy journey southwards for the next leg.

Mexico's event had a stuttering start, but in its current spell – which started only two years before Liberty Media's takeover of Formula 1 – it has finally flourished, showcasing the best of the city and the country, fuelled by a frenzied fan base who have rediscovered the sport.

23

SÃO PAULO GRAND PRIX: AUTÓDROMO JOSE CARLOS PACE, SÃO PAULO

> *'You feel the involvement around the circuit, it's crazy. The drivers, they love Interlagos, they love to drive it, the race is always one of the best races of the season.'*
> ALAN ADLER, CHIEF EXECUTIVE OFFICER,
> SÃO PAULO GRAND PRIX

Brazil is Formula 1's only stake in the ground in South America – and one of only two events located in the southern hemisphere; the other is Australia. Its grand prix takes place at the purpose-built Interlagos circuit, located on the southwestern edge of the sprawl of São Paulo, an expanding region that 22 million people call home. Here extreme wealth, displayed by the dozens of white skyscrapers and orange-roofed houses, intermingles with desperate poverty, grubby industrial quarters and rickety favelas.

Land acquired by developers in 1926 was initially designated to be converted into a resort, nestled between the lakes of Guarapiranga and Billings – hence the name of the circuit and the district: *Interlagos*, 'between the lakes'.

The economic crash of 1929 meant plans for a resort fell through and motor racing subsequently moved to the site in 1940, after a street race in São Paulo in 1936 was littered with issues. Formula 1 eventually arrived in 1972, initially as a non-championship event, and in 1973 Interlagos hosted the inaugural world championship Brazilian Grand Prix, when Brazil's Emerson Fittipaldi, the reigning World Champion, triumphed. The championship moved to the Jacarepaguá Circuit from 1981, in the home town of Nelson Piquet, but amid the rise and popularity of Paulista Ayrton Senna in the late 1980s, Formula 1 returned to Interlagos in 1990 on a truncated layout that manages to pack a lot into its small bowl-like space.

Brazil's race faced uncertainty in the late 2010s, heightened by political machinations between those in São Paulo and rivals in Rio de Janeiro – supported by the then President Jair Bolsonaro – who wished to move the grand prix to Deodoro, an area on the outskirts of their city. It would have meant the construction of a new facility, but there was doubt from the outset about the feasibility of such a project ever getting off the ground, while layers of environmental concerns – most notably the felling of a forest – added to the red tape.

Fortunately a new five-year deal was reached in late 2020 to preserve the race in São Paulo under a new promotional group, with support from Abu Dhabi's government. That replaced the organisation which had close ties and an arrangement with Bernie Ecclestone so favourable that the last events of the 2010s were not financially fruitful for Formula 1. (Ecclestone had departed as Formula 1 CEO in 2017.) As part of the restructure the event's name was officially altered to the São Paulo Grand Prix, reflecting the greater contribution made by the city – which owns the circuit – and state, and in 2023 a five-year extension was ratified to secure the event until 2030.

The news was well-received because Interlagos remains one of the best events on the schedule. The drivers adore the circuit, despite a relatively small paddock, with a narrow covered walkway and compact hospitality suites, a congested infield that means everything is slotted into tight spaces, and tricky access points into the facility, with a maze of avenues and alleyways, and a throng of fans sprawling across those routes.

The circuit is simple on paper, but it is relatively narrow, has some challenging and undulating medium-speed corners, and has a beautiful flow. At just 4.3km (2.6 miles), and with a lap taking only

70 seconds, it is among the shortest on the calendar and each mistake is magnified. Those who do err will find themselves taking a detour across the bumpy grass or will need extricating from the barriers.

'The track is iconic,' said Lewis Hamilton in 2023. 'They don't build circuits like this anymore. So it's a race that I think all the drivers really love. It's a race that it's fun to drive a single lap, but the races are very strong here. You've got that long straight which you can overtake on and the follow-through, Turns 1, 2 and 3 and then have a fight down to Turn 4, so a good overtaking place as well.'

The lengthy acceleration zone out of the uphill left-hander of Juncao and the full-throttle rise towards the pit straight have resulted in some scintillating battles to the finish line across the years.

'It's a place which will always remain special to me: 2019 was full of ups and downs, and it was my first podium,' Pierre Gasly says. 'Every single time I come back here, my friends always send me back the video of the side-by-side finish with Lewis and I feel a lot of positive energy!'

Brazil has a rich Formula 1 history, but there is one name that stands above all others when it comes to motor sport: Ayrton Senna.

His spectre casts a celebratory and commemorative shadow over both São Paulo and Interlagos. Even some 30 years after his untimely death, Senna remains one of Brazil's most known sons and his presence filters through the event and the city.

One of the highways near São Paulo's Guarulhos Airport – the terminus for international arrivals – is named after Senna, as is a tunnel in the centre of the city, while São Paulo's leafy Parque Ibirapuera has a monument and display in memory of the three-time champion.

In the vicinity of Interlagos, murals of Senna decorate some of the dilapidated buildings and creaking structures while spectators still don clothes and caps that represent Senna's unmistakable helmet of yellow with blue and green stripes, which was inspired by Brazil's national colours. Replicas of the blue cap advertising the now defunct Banco Nacional, iconically worn by Senna in his career, also remain a firm fan favourite.

Senna twice won Brazil's grand prix, in 1991 and 1993, and his first home victory is one of the championship's most legendary. Senna, by then already a two-time champion, overcame a failing gearbox in his McLaren to triumph on home soil and screamed in celebration on the cooldown lap at finally achieving victory in front of his supporters. He even required assistance to be extricated from the car, and to stand on the podium, such was the mental and physical energy he had poured into his drive.

At Interlagos a full-length painting of Senna from that 1991 podium ceremony dominates the wall of the paddock building by pit entry, stretching several storeys high, and is unmissable each time you drive into the track. An aluminium bust of Senna takes pride of place in the fan zone, while the opening downhill complex – where most of the overtaking happens during the race – is named the Senna S.

Drivers – even those born after his death – still honour his legacy, often wearing tribute helmets.

Senna's death in 1994 prompted three days of national mourning and he received a state funeral. His coffin was flown from Italy to Brazil, and wrapped in the green-and-gold flag. The funeral procession took place through São Paulo and an estimated three million people lined the streets to pay their respects, and the funeral itself was broadcast live on Brazilian TV. His final resting spot, Morumbi Cemetery, is often a place of pilgrimage for those wishing to pay tribute and features just a low-key plaque that is often surrounded by flowers and flags. On it reads the simple inscription: *Nada pode me separar do amor de Deus* (Nothing can separate me from the love of God).

Senna's foundation, set up by his sister Vivian after his death, continues to assist with education for underprivileged children throughout Brazil, funded by donations and licensed merchandise.

Senna's aura lingers over Interlagos, but the circuit is officially named after another Brazilian Formula 1 ace, Carlos Pace, who was killed in a plane crash in 1977 aged only 32. Pace, also a Paulista, claimed his sole grand prix victory at Interlagos, in 1975. Pace and Senna are just two icons of motor sport from a city and country rich in sport. The period after Senna's death was characterised by the failures of the fast but luckless Rubens Barrichello to win at Interlagos before, finally, Felipe Massa, his replacement at Ferrari, won in 2006. He doubled up in 2008, but that race remains ensconced in the memory for the astonishing culmination to the world championship, when Lewis Hamilton passed Timo Glock at the final corner to secure the fifth place he needed for the title. A dignified Massa celebrated amid the tears and it remains the last time a Brazilian driver won on home soil, while Barrichello's win a year later in Valencia is still the most recent victory for a Brazilian in Formula 1. Massa's retirement in 2017 left the country without a full-time representative on the grid for the first time in four decades.

Lewis Hamilton – who claimed a stunning wet-weather win in 2016 and surged from last to first in a spectacular drive in 2021 – was awarded honorary citizenship in 2022 and is hugely popular in the country.

'I won my first world championship here. It was kind of crazy back then, because I kind of felt like Public Enemy No. 1; obviously I was racing against Felipe,' said Hamilton during a press conference in 2023. 'But the growth that I've felt here, and the reception that I've had here, the amazing support. As a kid, growing up at home, being a fan of football, I always loved the Brazilian colours. And me and my brother would play FIFA, for example, and he would always take England, so I'd always take Brazil. And then there was Ayrton Senna, who was just such a big hero for so many of us. And when you come out here, you really feel his presence. And it's just such a beautiful culture.'

The passion for Formula 1 remains undimmed at Interlagos in spite of the absence of a local representative. In a city that is home to the likes of Corinthians and Palmeiras, there is a football-like atmosphere from the crowd, particularly in Grandstand A, which lines the lengthy banked section that leads to the start/finish line and overlooks much of the middle sector of the track. The seats at the front are mere metres from the circuit edge, allowing the fans to get up close and personal in a manner not possible at most other venues. Mechanics are acclaimed when they wander down the partially sunken and claustrophobic grid to wait for their cars to roll up, chants spring up for the most popular drivers and the carnival atmosphere continues for much of the day.

'I remember my first race here,' said Daniel Ricciardo. 'Like, the drivers' parade was cool. It felt like what you'd get at probably like a football match or something, you know, with lots of drums and flares and a little bit of dancing, and so that was cool.'

Grandstands lining much of the compact venue mean a cacophonous roar erupts for the start of the race, and for any side-by-side action or incident of note. The really hardcore fans wait in the arrivals of Guarulhos Airport for incoming drivers, even those landing bleary-eyed in the early hours of morning, while they line access route Avenida Interlagos at dawn and dusk to cheer on the participants. There is a deep-rooted and knowledgeable fan base, who are hugely interested in the intricacies of the sport, and Brazil is still one of Formula 1's largest television audiences.

'We are very emotional people, we cheer, we have this vibration, this culture,' says Alan Adler, Chief Executive Officer, São Paulo Grand Prix. 'The energy you find at Interlagos is really amazing. It's different to those new races where there's no tradition and there's not such a connection, and that translates to the place. You feel the involvement around the circuit, it's crazy. The drivers, they love Interlagos, they love to drive it, the race is always one of the best races of the season. If you put all that together, it can explain the success of Interlagos.'

That heightens the experience too for Formula 1 personnel during a draining stage of the campaign amid the continuing expansion of the schedule.

Brazil used to be one of the few 'flyaway' events when the season was shorter and the number of grands prix was in the high teens, and until 2003 it featured as one of the first rounds of the year. From 2004 through to the early 2010s it was regularly the season finale, leading to a tense atmosphere and some dramatic showdowns.

Fernando Alonso clinched both of his world championships, in 2005 and 2006, at Interlagos. 'I think Brazil will always be good memories for me,' he said. 'Every time I come here, I remember the championships, I remember some nice battles in the past. I love the circuit. It has a good combination of corners, a lot of action always here in Brazil.'

Kimi Räikkönen sealed his 2007 title at the Interlagos finale, Hamilton secured a last-gasp triumph in 2008, and in 2009 it was Jenson Button's turn to be crowned at the venue.

'I remember 2015, the first time I travelled to São Paulo for a Formula 1 race and getting to race on this iconic track,' says Carlos Sainz. 'I always remember like, São Paulo, like, it's happening. I played so many times on the PlayStation, but I remember watching all the world titles being decided here when I was a kid, and I saw Lewis winning on the last lap, Fernando winning his first world championship. It's always a track that I wanted to come to and I remember 2015, the first time I came, as a special weekend.'

The evolving calendar has meant Brazil has shifted away from being a title decider, or even a standalone voyage, and now regularly comes at the end of a punishing triple-header that includes Austin and Mexico, and before a final flourish of rounds elsewhere. It means that personnel can be drained by a third week on the road and media attendance is quite low. Getting around is not the most straightforward owing to the city's congested routes and insufficient infrastructure. It's not quite as bad as Mexico City, but journeys in São Paulo can be laborious, particularly on Friday evenings as commuters criss-cross the city's congested roads amid a sea of red brake lights. Traffic jams are even commonplace inside the tight confines of the facility and the brief drive from the car park to the main gate is known to take up to an hour. Alternative means are available, most notably the subway system, which has a stop near the back end of the circuit. Two-wheel transport is commonplace, the mopeds and motorbikes weaving in between the bottlenecks at speed, honking as they duck through perilously small gaps to alert motorists of their presence. The elite look to the skies, opting for helicopters, and São Paulo is the only city in the world with its own air traffic control team exclusively for the choppers. It means the business districts in which Formula 1 personnel tend to stay often hum along to the sound of the airborne creatures.

There is also an edginess to the journeys through the chaotic roads. Personnel travelling back from the circuit have been accosted at gunpoint, so precautions must be taken and there is an extra diligence.

That doesn't mean that a weekend in São Paulo is a case of *lock the hotel door and hide away* – far from it. No trip to São Paulo is complete without a detour to one of the churrascarias (barbecues) of Morumbi and Berrini, and most evenings these almost resemble the Formula 1 paddock. The most popular is Fogo de Chão, a restaurant chain that specialises in the traditional gaucho method of roasting meat over an open fire. There is a set price and waiters zip around the restaurant holding long skewers, presenting and slicing the different cuts of meat, which you then pluck with provided tongs. Table mats coloured green and red indicate whether you still require further cuts, though the servings will probably keep coming. There are also plenty of bars and clubs throughout São Paulo where the traditional (and dangerously strong) Caipirinha cocktail is served. Another delicacy gorged on by the paddock are the pão de queijo, which are small baked cheese balls with a chewy exterior and elastic soft interior, and which are allegedly meant for breakfast but permissible at all times of day. They are available at a stand in the paddock and, unsurprisingly, they vanish as soon as they emerge from the oven.

Interlagos is also known for its capricious weather. Whole days can be spent basking in the sunshine, but shower systems can crop up out of nowhere and almost slip through the weather radar, meaning teams and drivers must be quick-thinking. In fact, the signal for the change in conditions is often a glance out into the uncovered spectator areas to see ponchos being donned and umbrellas erected. Showers heightened the drama in the dramatic denouements to the 2008 and 2012 seasons, and changeable conditions have facilitated shock qualifying results, most notably in 2010 when a rookie Nico Hülkenberg scored pole, or in 2022 when Kevin Magnussen and Haas stunned by setting the fastest time.

'From the beginning of my career I somehow clicked with this place and track. Always love driving here,' says Hülkenberg. 'Just the rhythm, the elevation changes are quite cool, the twisty infield stuff versus the faster other sectors. It's a fun place. And just, you know, the heritage and history, I think you can feel that here and I embrace that and like that.'

There are also days where the rain can simply deluge Interlagos. Wet races in 2003 and 2016 were littered with incidents, accidents, spectacular saves and stoppages, as the track surface took inspiration from the nearby reservoirs.

In 2003 over half of the field were out by the time Mark Webber crashed heavily and Fernando Alonso piled into the debris. Only 56 of 71 laps had been completed, but the race was called and the results backdated two laps per the regulations, putting Kimi Räikkönen victorious ahead of Giancarlo Fisichella – who had taken the lead just beforehand – and a hospitalised Alonso third amid much confusion. The podium ceremony therefore featured only two drivers – neither, it later transpired, in the correct order. After a timekeeping error was identified, it became clear that Fisichella had taken the lead in the sufficient window – per Formula 1's slightly convoluted regulations – before the race was stopped, and he was installed as the winner, his first victory in Formula 1.

The 2016 race featured a lengthy red flag due to the conditions. Lewis Hamilton triumphed, while Max Verstappen reminded any doubters about his potential after a phenomenal mid-race save and a surge from 16th to third across the final 15 laps.

'2016 maybe for me wasn't like a highlight result, just that race was wild,' said Daniel Ricciardo. 'I remember that as a crazy one. I think Kimi [Räikkönen] was stuck on the straight and there was a red flag or something. I didn't even see Kimi. I didn't know what the red flag was for. So visibility was just insane. Yeah, foggy visor. It was just wild. So that was a chaotic race!'

Brazil remains Formula 1's only outpost in South America, a continent with a passionate fan base, but one broadly without the infrastructure, available finance or desire to host a grand prix. Argentina, home nation of five-time World Champion Juan Manuel Fangio, hosted 20 grands prix but dropped from the calendar permanently after 1998 owing to a lack of money. Post-pandemic, Colombia's coastal city of Barranquilla emerged as a possible location for a grand prix. Formula 1 CEO Stefano Domenicali visited the city in late 2022, but discussions did not progress beyond the initial phase and the project quietly disappeared.

São Paulo and Interlagos rarely disappoints. It is only a shame that the expansion of the schedule has taken away some of the shine from an iconic event, which falls nowadays at the end of an arduous American leg.

Interlagos, nestled within the sprawling Sao Paulo, is an iconic venue

24

LAS VEGAS GRAND PRIX: LAS VEGAS STRIP CIRCUIT

'When I first saw the layout, I was like, Oof, this is going to be a boring track. *But then when I actually [got] here . . . it felt good. I think it's a good track.'*

CHARLES LECLERC

In a city where there's always bombast, extravagance and excess, Formula 1 found a new and shiny home, slicing past the casinos and resorts along the world-famous Strip, and putting on a show in Sin City.

Formula 1 burst on to the streets of Las Vegas in 2023, the event receiving enormous publicity in the build-up, and became the third F1 annual event on US soil.

This was not the first time Formula 1 had headed to Nevada's Las Vegas, but the prior two visits in the early 1980s proved to be failures.

A temporary circuit was laid out in the large open-air car park of the hotel of Caesars Palace, which was backing the race, but the layout was flat, featureless and repetitive – and the outskirts of Las Vegas made for a backdrop that was drab, flat and featureless. The races were held in the afternoon, with extreme heat in 1982 nudging towards 40°C (104°F) and adding to the discomfort for spectators and drivers. The event lacked the required widespread support of the city officials and influential figures, and it was also poorly attended; the 1982 race is estimated to have welcomed only 40,000 punters.

Las Vegas – or rather, Caesars Palace – went down in ignominy and 'Caesars Palace' entered F1 parlance as a euphemism for failure.

But as Formula 1 continued its expansion in the United States, and with the Miami Grand Prix project underway, the championship began seriously exploring whether a race in the heart of Las Vegas was a viable proposition. Las Vegas was interested, if not dramatically so, but the landscape began to evolve.

'I originally talked to Chase Carey in 2019 and then the pandemic happened,' says Steve Hill, CEO of the Las Vegas Convention and Visitors Authority. 'I started to watch *Drive to Survive* as kind of homework. I felt like I needed to learn more about this sport, and I saw what was there and thought, *Wow, this is something*. I went to Austin, in 2021, and you see what that does, it overwhelms Austin, it's crazy, and it's out in a field and it's still such a big event. You see it there and see what that can be to Las Vegas, and you think, *This is going to be something*.'

Hill admits he 'thought it'd be hard and logistically difficult' to put Formula 1 into the heart of Las Vegas, because he 'couldn't see whether the hotel resorts would see it would work.'

Those hotel resorts have substantial influence in Las Vegas. Crucial to the agreement was being able to bring Formula 1 into the very centre of Las Vegas. A track on the edge of the city, or at its permanent Las Vegas Motor Speedway, would have been a case of *So what?* and Formula 1 knew it had to secure the ability to race along the Las Vegas Boulevard, better known as the Strip.

The Strip is the focal point of Las Vegas, an ecosystem dominated by bars, restaurants, theatres, nightclubs, more Elvises than you can shake a stick at, and most prominently the hotels, casinos and shopping precincts that have their own enormous infrastructure, attracting high rollers and tourists in equal measure. The Strip has grown as the centre of Las Vegas as a destination city across the decades and now five of the world's ten biggest hotels are located along the Strip. The hotels, casinos and shopping malls intermingle, the lobbies and check-in desks a stone's throw from the reams of slot machines – the main hotels have over a thousand each. Restaurants endorsed by celebrity chefs come with views of roulette wheels and blackjack tables, and the entrances to theatres where world-famous artists have their residencies are navigable through the craps tables and video poker machines. The relentless noise, booming music, dizzyingly patterned worn carpets, the flashing lights and the stench of cigarette smoke, as well as the absence of natural light, enhance the temporal distortion, and the outside world may as well just be a construct. It is an ostentatious near-24/7 fantasyland where the best and worst of Las Vegas are on display, and there is nowhere else quite like it. And just off the Strip it is a different world: low-key, dingy and occasionally a little bit grubby.

Another crucial step was Formula 1, through owner Liberty Media, taking on responsibility – and the financial risk – as the promoter of the grand prix, the first time that the championship had taken direct involvement.

'We chose to do it ourselves largely because it's close to our home in Colorado, it's a place where we thought we could work with the partners, a place where we would not have a need to work with a foreign government, and a place where we could cut deals, and build a site and invest,' Liberty Media President Greg Maffei said in 2024. 'We did invest, $650 million between what we bought and the property, and what we put into it. That's a scale investment that, I think, tops any other investment that someone's put into a site for a Formula 1 race.'

It gave the major hotel chains and casinos – including the likes of the MGM Grand, the Wynn and the Venetian with its 7000 rooms – as well as government authorities at various levels, the assurance that Formula 1 was serious about Las Vegas and in it for the long term, rather than seeking to exploit the city as a swift get-rich concept. Las Vegas did not need Formula 1 as a promotional tool – it exists in its own self-fulfilling bubble – but Formula 1 dearly wanted Las Vegas, given its reputation and status as the self-proclaimed entertainment capital of the world.

An agreement was reached in March 2022 for a three-year deal to enable Formula 1 to compete in Las Vegas, receiving the sign-off from the Las Vegas Convention and Visitors Authority, and the support of Caesars Entertainment, MGM Resorts International and Wynn Las Vegas, known as the Founding Partners.

Before the inaugural grand prix took place, Nevada's Clark County approved a resolution for Formula 1 to hold the Las Vegas Grand Prix on the weekend before Thanksgiving until at least 2032, should there be the desire to do so. As part of the arrangement the race was installed on Saturday, marking Formula 1's first non-Sunday race since South Africa in 1985. The rest of the weekend's programme moved up a day apiece, with practice taking place on Thursday and qualifying on Friday. That ensured Formula 1 got its night race, while the time zone difference meant it was a Sunday morning race in its European heartlands. For fans on America's Eastern seaboard, though – well, unlucky.

'We anticipate a lifetime together in partnership,' said Clark County Commissioner James Gibson upon the unanimous approval of the motion.

Formula 1 was able to use a 1.9-km (1.2-mile) stretch of the Strip, ensuring organisers had the showpiece focal point they craved, but the thorny issue of the pits/paddock complex remained unanswered. There was speculation that a temporary structure might be required, with equipment transported to and from a paddock a mile away, but this was swiftly dismissed when Formula 1 opted to acquire 16 ha (39 acres) of land off East Harmon Avenue and Koval Lane, a couple of blocks east of the Strip. That was another indication of Formula 1's long-term interest in Las Vegas: the land cost $240 million and another hefty chunk – stretching into the hundreds of millions – was set aside to construct a permanent three-storey pit building. It was an expensive undertaking, and costs rose due to inflation and the brief timescale involved, but it put Formula 1 in charge of its own destiny and provided a year-round presence in the heart of the city. It meant Formula 1 was able to construct a building to its own specifications, while also maximising paddock space and – perhaps more crucially – installing large hospitality areas for its premium and lucrative Paddock Club, where tickets start at $3750 and the higher-end packages soar into five or even six figures.

Formula 1's pit building, structured across three levels and around 300m (980ft) long, was the latest sporting structure in a city that is increasingly developing into a sports hub. The T-Mobile arena, which has hosted big boxing fights, opened in 2016 and four years later the multi-purpose Allegiant Stadium was finished. It is primarily the home of NFL team Las Vegas Raiders, which relocated to the city from Oakland in 2020, and it hosted Super Bowl LVIII in 2024. A new stadium is under construction, set to open in 2028, when MLB team Oakland Athletics will relocate to Las Vegas. All these stadia are within a stone's throw of the neon-filled Strip.

Formula 1 had the Strip secured and the pit building location sorted, but nonetheless still needed an actual circuit. Tilke Engineers and Architects – which has had a hand in a catalogue of new-build circuits since the turn of the millennium – was recruited to assist with the design.

'The main goal for everyone was to use the main features of Las Vegas,' explained project leader Carsten Tilke in an interview in 2023. The Strip accounts for almost a third of the circuit's length, but there is another full-throttle section along East Harmon Avenue, a long straight along Koval Lane, and slow-speed turns at the end of the Strip and around the Sphere.

'From that starting point we looked at different configurations and layouts – and the next question is always: *Where do you put the start/finish line? Where do you put the garages, the team hospitalities?* says Tilke. 'And then Liberty Media had the opportunity to buy the piece of land [for the paddock] and that was the missing piece.'

The street-based nature means 'you are very limited in the space that you have, and you cannot build completely new streets or go over parking lots,' but there are still factors to consider: how to make a circuit that will be enjoyed by drivers, be good for racing, provide a good spectacle for TV audiences and trackside spectators, while also limiting disruption for the city and ensuring hotels, shops and facilities within the track still have access.

'We had a good phrase,' says Tilke. 'Which was: *It's one piece art, it's one piece science and it's one piece magic!*

'So you have the Strip, then you have an area where you're doing the paddock, and then of course how to connect these two areas in a track length, which is about 4–6km [2.5–3.7 miles], maybe

7km [4.3 miles]. You don't want to have it too long, nor too short. So then you look at what streets can you use, what is actually the possibility of streets you can use and, if you know the streets which you can use, how can you make this as exciting as possible for the drivers and where can you put spectators, and also make nice areas for spectators.'

Considering the geographical limitations and the compromises that had to be made to connect various areas, the circuit received a largely positive reception. Organisers also utilised the presence of the newly constructed Sphere, an entertainment venue that has 1.2 million tiny LED screens on its exterior and which has already established itself as a Las Vegas landmark.

'Street tracks are my favourites,' says Charles Leclerc, who claimed pole position for the opening Las Vegas race. 'When I first saw the layout, I was like, *Oof, this is going to be a boring track*. But then when I actually [got] here, from the first laps on the simulator it felt good. I think it's a good track.'

Teams have sophisticated state-of-the-art simulators inside their headquarters where drivers can practise virtually, with each track scanned and modelled to ensure as accurate a representation as possible.

'It's been quite surreal being here and seeing everything that that's going on,' says George Russell. 'The track, it was actually a lot better to drive than I anticipated. It looks pretty basic from the track map, but it's actually got quite a lot of character, really challenging circuit to drive.'

'I love Vegas,' says Logan Sargeant. 'I used to race here in karts. It always went really, really well. And it's nice to come back taking an F1 car down the Strip!'

Las Vegas is Formula 1's youngest event and inevitably there have been teething troubles and complaints – some of which more justified than others. The bold and brash approach by organisers, with phenomenal hype and positivity relentlessly drummed into observers, has done little to assuage critics.

The lengthy section of the Strip used by Formula 1 was resurfaced in advance and the construction work caused months of disruption to a key area of Las Vegas, prompting understandable frustration from locals and tourists alike. The circuit build then also took time, causing additional disruption, while temporary structures such as bridges negatively impacted businesses that were bypassed. The removal of trees next to the Bellagio Fountains, and the famous water display being blocked by a grandstand, also came in for criticism. There were more than a handful of comments from irascible locals implying that Formula 1 needs Las Vegas more than Las Vegas needs Formula 1.

'In Monaco, I can definitely see since I was very young the whole organisation to get to the race day and for everything to be ready and how much effort is put into it,' Charles Leclerc explained. 'But we have that since such a long time in Monaco that I feel like everybody got used to this preparation and also kind of enjoy now the moment and obviously the event as a whole. It's obviously very new for Vegas and for the people in Vegas, so I can understand that it's difficult to accept at first, but I really hope that they saw the benefits of having so many people coming to the event.'

The weekend's timetable also came in for criticism.

Most grands prix start at 14:00 or 15:00 local time, twilight events at 17:00 or 18:00, and night races usually at 20:00.

Holding the Las Vegas Grand Prix as a night race was essential for a multitude of reasons. Las Vegas' prime-time events typically take place late in the evening and closing the streets off after dark – and reopening them prior to first light – minimises disruption for the city and its residents. What's more, Las Vegas is a spectacle at its best at night.

But a race start time of 22:00, with qualifying at 00:00, only adds to the sense of distortion for personnel.

Unlike heading to Singapore, which means travelling eastwards and staying on European time, travelling to Las Vegas incorporates a time zone shift of eight hours to Pacific Standard Time (PST), followed by switching to an unnatural night-based schedule, which effectively means adapting to a Japanese time zone. And unlike at most night events – which have a warm climate – the temperature in Las Vegas in late November after dark is chilly, with sunset at an early 16:30, meaning personnel are wrapped up in winter coats and beanies, huddling around heaters and barely seeing any daylight all weekend. That factor was most prominent during an inauspicious start to the inaugural grand prix in 2023.

A loose drain cover along the Strip was struck by Carlos Sainz after only eight minutes of practice, which wrecked the floor of his Ferrari – a costly undertaking and one from which Sainz was fortunate not to incur injury – and no running was able take place until everything was checked and deemed safe. That was only complete after several hours and consequently practice resumed at 02:30, concluding at 04:00. Even by Las Vegas' standards as a day-round city, this was surreal, exacerbated by fans being evicted from the site at 01:30 owing to the hours that security and transportation staff could legally remain on duty. That meant Formula 1 cars were hurtling through the city at an absurd hour of the day and with no one to watch them. A smattering of resourceful spectators found vantage points by going up and down the escalators adjacent to some of the bridges that traverse the track. Fans affected were offered a $200 merchandise voucher – paltry considering the exorbitant prices – and a half-hearted non-apology the next day. Formula 1 learned the hard way that race promotion is not a walk in the park.

The schedule was universally loathed, from drivers to teams and spectators, and for those on the East Coast of America it meant a race start time of 1 a.m.

'The only two things I would critique would be the asphalt, we need to get some more grip out of it, and just the start times, bring them earlier and that would be a little smoother,' said Daniel Ricciardo after the inaugural event.

Charles Leclerc labelled the late schedule as 'a bit on the limit', with mechanics clocking out of the paddock after first light and returning to bed once the sun was high in the sky. For some, there was a further blow. A handful of hotels, particularly those with views of focal areas, were mandated to perform daily security sweeps from 08:00, checking rooms for weapons. That did not go down well with those on an irregular sleep pattern just trying to work.

The Las Vegas Grand Prix also brought the *sport versus entertainment* topic to the fore.

As a high-profile event there is a greater focus than usual on 'the show', and an increased demand on drivers, who are pulled this way and that. The joke goes that drivers are not paid for racing, but for everything else they do outside the car, but it is always pertinent to remember that they are humans and not performing monkeys at the zoo, particularly when they're trying to get in the zone before competition.

'A kind of show element is important, but I like emotion and for me, when I was a little kid, it was about the emotion of the sport, what I fell in love with, and not the show of the sport around it because, as a real racer, that shouldn't really matter,' said Max Verstappen during the 2023 event, part of several monologues on the topic, in which he suggested Monaco was 'Champions League' and Las Vegas 'National League'.

The argument is likely to bubble on for several years, but once Las Vegas' grand prix matures across the following years it can begin to build a legacy that some tracks have been gathering for decades.

Las Vegas nonetheless proved in 2023 that sport and entertainment are not mutually incompatible. There were several music shows, an abundance of off-track activations and tweaked showbiz-related elements, but come Saturday night Formula 1 delivered one of the best races of the season around Las Vegas' streets. Verstappen was victorious while dressed in an Elvis-themed race suit, there were battles up and down the grid, second place was settled only on the last lap, and there were incidents and clashes aplenty too.

Formula 1 has a vested interest in ensuring Las Vegas is as successful as possible and there is little doubt that the grand prix has already emerged as a showpiece event. There were setbacks, and justified criticism, but it's always important to take a step back and look at the bigger picture. Getting Formula 1 cars to race down the Strip – one of the most famous avenues in the world – was an *unthinkable* proposition just a decade before. That this is now on the schedule highlights the championship's evolution and its greater affinity with the United States under Liberty Media.

The city would feel suitable as an end-of-year season finale location, given the pizzazz, the hype and the spectacle, but there's a couple of lucrative blasts in the Gulf yet to run.

25

QATAR GRAND PRIX: LUSAIL INTERNATIONAL CIRCUIT, LUSAIL

'I do like the high-speed section, so coming to Turns 14, 15, it's almost flat out. It's just on the edge of being flat out, so it's really a corner where you are really maximising the load of the car and the Gs you're feeling there.'

PIERRE GASLY

The shortest distance between tracks on the Formula 1 calendar is between two events that are at almost opposing ends of the schedule.

In 2004 Formula 1 established itself at the Bahrain International Circuit. Later the same year another venue in the region was opened courtesy of the foundation of the Lusail International Circuit, just outside Qatar's capital city Doha. Only 110km (70 miles) and the Persian Gulf separate the two venues.

Initially Lusail was the home in the region for the MotoGP world championship. Not until 2021 did Formula 1 set foot in Qatar and that came at relatively short notice. The pandemic and consequent travel restrictions kept several countries, most problematically Australia, Singapore and Japan, off limits to Formula 1, and as it juggled its schedule Qatar came on the radar: the facility complies with the FIA's mandatory Grade 1 standard and a swift arrangement was reached. It was just five weeks between the announcement and Formula 1 cars emerging on to the circuit at Lusail.

This was effectively Qatar dipping a toe into the water because by the time it hosted the 2021 grand prix it had already penned a 10-year deal to join the schedule from 2023. Why not 2022? Well, the relatively small nation was devoting its resources to hosting the most compact FIFA World Cup in history, the stadia all condensed into the capital Doha and the surrounding districts. Qatar invested heavily in the tournament, not just in the stadia, but on infrastructure and transport connections in and around Doha. It used the World Cup as a tool to bolster its reputation on the global stage and successfully completed a game of regional one-upmanship. On 18 December, 2022, Lionel Messi, then employed at club level by the Qatar-owned Paris Saint-Germain, wrapped in a bisht (the traditional men's cloak), hoisted aloft the World Cup trophy in the Lusail Stadium, watched by over a billion people worldwide.

The World Cup was done. Enter Formula 1.

'We are the new World Cup for the next 10 years,' said Amro Al-Hamad, Chief Executive of the Qatar Motor and Motorcycle Federation during a press conference in 2023, which was held on a yacht in Monaco. 'Every year we'll have to improve and develop an event that is suitable for the name of the state of Qatar. We want to see Qatari students and new youth being part of this vision. We do expect to see Qatari engineers and designers working with F1. It's not only focused on having a race over 10 years.'

It also helps that Qatar has money. And lots of it. It may not have quite the oil reserves of sole land neighbour Saudi Arabia, but it has more than enough to be able to attract major players to its peninsula. Qatar's return to the calendar as a permanent fixture also coincided with national airline Qatar Airways replacing Dubai's Emirates as one of Formula 1's major partners, striving to expand its reach in the sports and travel landscape.

Qatari organisers considered establishing a street-based venue in downtown Doha, but the idea was short-lived and instead its sole permanent motor-sports facility at Lusail was heavily renovated. The circuit layout was unchanged because officials were keen to preserve long-standing MotoGP track records – though it got a resurface and a brush-up – and the remainder of the complex was substantially overhauled.

Lusail, located around 15km (9 miles) north of Doha on the eastern side of the peninsula, rises from the desolate desert at the moment, but given the region's propensity for rapid and vast construction it will not be a surprise if Lusail is eventually swallowed up by expansion. The highway, at points up to eight lanes wide, passes the enormous gold bowl-like Lusail Stadium, and has overhead gantries directing spectators, media, VIPs and VVIPs to those stadia. Imagine being a VIP and discovering that there is still another level and, you suspect, a further level for those not wishing to advertise their presence.

A completely new paddock, medical centre, media centre, race control, spectator areas and accompanying access roads, routes and car parks were constructed in advance of Formula 1's return in 2023, making the facility almost unrecognisable compared to 2021. This was a classic case of regional competition to trump its counterparts in Saudi Arabia, Bahrain and Abu Dhabi. That said, it was not without imperfections, at least in its first return year. A few buildings remained incomplete internally, enormous car parks were clearly built with an improbably large number of attendees in mind, while security had not quite been briefed on some of the access requirements for personnel. Somehow, no one in the media centre could locate the climate control remote, so it was affixed on the coldest setting all weekend.

Lusail was not necessarily designed with Formula 1 in mind, but it has emerged as a fiercely fast layout. There are 16 turns across the 5.4km (3.3 miles) of its layout and only a handful of them require drivers to dip below 100km/h (62mph), with fourth gear and below used only sparingly. Most of the corners are medium to high speed and several of them are high G-force, long-radius corners in which drivers spend several seconds.

'I do feel this track has quite a lot of character,' says Pierre Gasly. 'It's very twisty. I do like the high-speed section, so coming to Turns 14, 15, it's almost flat out. It's just on the edge of being flat out, so it's really a corner where you are really maximising the load of the car and the Gs you're feeling there.'

Lewis Hamilton triumphed upon Qatar's debut in 2021 and when it returned two years later Max Verstappen clinched his third world title at the grand prix, giving a nascent event a slice of Formula 1 history.

Track sessions taking place at different times of day can pose set-up challenges owing to different surface temperatures, while the wind can also be erratic and violent at Lusail, blowing across a vast expanse before reaching the circuit. It is punishing for the drivers and punishing for the tyres. Both of these elements have manifested themselves in Qatar.

The only contact patch connecting the cars to the track surface is the tyres. Formula 1 tyres have evolved through the decades, with different suppliers and various requirements, but since 2007 there has been only a single provider, when Michelin's withdrawal left Bridgestone alone. This helped Formula 1 with cost control and that approach has been retained, Pirelli replacing Bridgestone for 2011 and remaining the sole supplier through every subsequent contract cycle. The company is also one of Formula 1's main partners, having a commercial presence, and sponsors several grands prix.

Pirelli's products have evolved through its stint as tyre supplier and it currently has five dry-weather compounds, from which three are taken to each grand prix depending on the circuit characteristics. The Hard compound lasts the longest and has a white band around the sidewall; the red-banded Soft compound is the quickest across a single lap and has the least durability; and the yellow-banded Medium compound sits in the middle. Drivers have 13 sets of dry-weather tyres available for each grand prix weekend and must use Soft tyres in the final part of qualifying, and two different compounds in a race if it takes place in fully dry conditions.

Pirelli faces an onerous task. It must produce tyres that can cope with the performance levels of 10 different, ever-evolving Formula 1 cars, and that are suitable for 24 different circuits with various surfaces and corner types, while ensuring there are time deltas and degradation levels between compounds that result in varying strategies for attractive but not fabricated racing. A total of 6847 sets of tyres were used by drivers through the course of the 2023 season.

'Formula 1 is the top of the requirement of technology,' said Marco Tronchetti Provera, CEO of Pirelli, when announcing the company's 2025–27 contract. 'Speed and G-forces in Formula 1 are amazing; we cannot reproduce this in any other kind of motor sport. Motor sport for us is a priority in general to provide extreme conditions for testing, because for us motor sport is an open air laboratory. Within this lab Formula 1 is the top level in which we test new materials, but where every weekend when looking from outside it seems the same, but every weekend the cars are different. We have to set tyres that are able to fit each driver in the same way, or each car in different conditions, and different weather conditions, but also the [surface] is different. We have to continue being a leader and Formula 1 is helping us be a leader.'

Pirelli gains from its presence in Formula 1, but the widespread publicity it receives is usually restricted to when something goes wrong, as it has done in Qatar. In 2021 a handful of drivers suffered front-left punctures during the late stages of the race, and after practice in 2023 Pirelli engineers detected micro-cuts on some tyres, indicating the potential for a problem. They pointed to Lusail's aggressive kerbs, the high speed and the prolonged time drivers spend on them as the likely culprit. Remedies were implemented, which included a limit on stint lengths during the race, with drivers permitted no more than 18 laps on one set of tyres. That meant they were able to push flat out at a circuit with fast and relentless curves – and accentuated another factor drivers have had to contend with in Qatar: the oven-like heat.

Scheduling 2023's comeback event in early October resulted in daytime temperatures of up to 42°C (107°F) – the hottest encountered all year – and while the main track activity took place in cooler night-time temperatures, the humidity levels can vary and on bad days can exceed 80 per cent. That makes for an especially unpleasant experience for everyone – with moisture immediately forming on surfaces outside – but particularly for those in the cockpit. A couple of drivers had trouble extricating themselves from their cars post-race and many conceded they found Qatar the toughest race they had ever experienced, reporting illness, light-headedness and blurred vision. This all came as a surprise: very few drivers were braced for any grand prix to surpass Singapore for heat and humidity.

'We probably found the limit,' said Lando Norris. 'I think it's sad we had to find it this way. It's never a nice situation to be in, you know, some people ending up in the medical centre or passing out, things like that. It's not a point where you can just go: *The drivers need to train more* or do any of that: we're in a closed car that gets extremely hot in a very physical race.'

Max Verstappen, who won that race, described it as 'just too warm, and it has nothing to do with more training or whatever. I think some of the guys who were struggling are extremely fit, probably even fitter than me, but just the whole day it's like you're walking around in a sauna and then in the night the humidity goes up.'

Some drivers tried opening visors at stages, only to be buffeted by sand and dust, while others attempted to funnel air into their overalls through their sleeves along the straights.

Formula 1's governing body, the FIA, began an investigation and approved the use of cockpit 'scoops' in extreme weather, effectively allowing air to be channelled to the driver. But Qatar's shift in the schedule from 2024, from early October to late November/early December when the weather is slightly cooler, is expected to ameliorate the issue in the long term. There was, after all, a reason FIFA opted to grant Qatar the first winter World Cup.

After the boisterous and popular American leg, Qatar is at the opposite end of the spectrum. The country is still emerging on the main stage, there is very little pizzazz around the event and spectator attendance is among the lowest of the season. A weekend figure of 120,000 was released for 2023 – around a quarter of Formula 1's most popular events – and even that number was greeted with scepticism by those staring at sparse grandstands.

The majority of personnel set up home for the weekend in Doha's West Bay. This is a spit of land perched on the northern side of the capital's crescent-shaped corniche, which stretches for 7km (4.3 miles). As recently as the early 1980s the corniche, and West

Bay, were just barren, sandy expanses, completely unrecognisable from their current existence. Today, this is where several cultural landmarks are located and it is now an orderly district full of high-rise, glass-fronted international hotel chains, resorts and shopping malls, government ministries and international embassies, and cafés, restaurants, clubs and fast-food chains imported from elsewhere. West Bay is best at night, when the colourfully illuminated skyscrapers are reflected in the shimmering water, on which bobbing wooden dhows provide a reminder of Qatar's past as it seeks to accelerate its future.

More authentic, in the heart of Doha, is the pedestrianised Souq Waqif, with labyrinthine covered alleyways of market stalls, shops and cafés, where vendors are enthusiastic and eager, shorn of the unpleasant hassling and aggression of counterparts in other countries. It is a blend of the historic and the modern, having been refurbished and renovated according to traditional architectural designs in the mid 2000s after a fire, and after having fallen out of fashion in the 1990s. The sand-coloured stone buildings, with exposed wooden beams and bamboo poles from which lanterns swing, display their products outside and entice the shopper inside to shelves crammed with options. There are rows upon rows of shops with exquisitely embroidered colourful abayas and hand-knitted patterned carpets; around the corner are local spices, herbs and perfumes that create a cocktail of exotic scents; and on the next alley are crafted pieces of glistening gold jewellery. The bustling marketplace has been designed to feel like a step back in time, providing the essentials for residents and a glimpse into Qatari life for the curious visitor, who are likelier to be less at ease with the caged birds, other animals and tied camels also present at the souq. And yet, there are endless gift shop souvenirs, discarded World Cup memorabilia and occasional Formula 1 merchandise dotted around too, and beneath the souq is a multilevel air-conditioned car park. There is also, inexplicably, a golden statue of a giant thumb plonked in the restaurant district, while an enormous square filled with benches is empty in the scorching heat of the sun, but becomes an energetic meeting place in the evenings.

Qatar lacks the blingy tourism sector the United Arab Emirates has astutely exploited, is behind Bahrain in terms of Formula 1 history, and is trailing Saudi Arabia when it comes to sports leverage, but the peninsula is gradually becoming a bigger player in every metric on the Gulf stage. It is a race that is largely a case of *tick the box, get through it and move on*, as the arduous season trundles along with different race timetables and conditions. But, as with other money-laden venues, such as Saudi Arabia, Formula 1 is expected to be a large part of Qatar's promotion in a global arena long into the 2030s and beyond, so it is here to stay.

Traditional dhows bob on Doha Bay, in front of its rapidly expanding business district

26

ABU DHABI GRAND PRIX: YAS MARINA CIRCUIT, ABU DHABI

> *'They changed the track layout a couple of years ago… it's made the driving experience a little better than before, [but] I don't feel it has affected overtaking opportunities, so it hasn't improved the actual racing.'*
>
> DANIEL RICCIARDO

Formula 1's long and arduous championship reaches its final terminus in the United Arab Emirates in early to mid-December, the sun setting on the season at Abu Dhabi's twilight race.

As with Formula 1's other Gulf partners, Abu Dhabi is a lucrative ally for the championship. Despite only debuting in 2009, Yas Marina has already held more season finales in the championship's 75-year history than any other circuit, now over a dozen, and is contracted through to 2030. The grand prix weekend regularly coincides with the UAE National Day on 2 December, meaning the region is decorated with flags, banners and lights, reflecting the number of years since its 1971 independence.

The Yas Marina Circuit is part of an ever-expanding sports and entertainment hub on the man-made Yas Island. It is located around 30 minutes from downtown Abu Dhabi, the slightly less famous of the United Arab Emirates' main two emirates, and the ever-expanding Dubai is 90 minutes in the other direction along the highway that slices through the flat lifeless desert. As with several other Gulf states – Saudi Arabia, Qatar, Bahrain – the UAE uses sport as a promotional tool and for influence. Its most notable sporting arm is its ownership of Manchester City since 2008, and airlines Etihad (Abu Dhabi) and Emirates (Dubai) are prominent sponsors across sports and entertainment disciplines.

Across from the circuit's main entrance is Yas Plaza – a complex of hotels that circles a piazza where the majority of paddock personnel stay – while further around from Yas Marina Circuit is the enormous Yas Mall, full of western boutiques and connected to Ferrari World. There are also Yas Waterworld, Seaworld and Warner Bros. World, complemented by a range of still-rising hotels and apartment blocks, creating a synthetic landscape as Abu Dhabi attempts to transition towards tourism in the long term. *Throw money at it and they will come* – that, at least, is the hope. Downtown Abu Dhabi has more authenticity, aided by the blending of cultures – usually influenced by the workers who have migrated from the Indian subcontinent – but that's not where Formula 1 rocks up for its finale.

'It's a nice place to finish the year,' says Oscar Piastri. 'Nice and warm – it's halfway back to Australia.'

The mood in the paddock at the end of the season is very much influenced by the situation in the championship.

If the world title is up for grabs, there is an edge to the atmosphere, the nervous anticipation simmering throughout the weekend, and an extra weight of importance placed upon each on-track session as the paddock counts down towards Sunday evening, when the destination of the championship will be settled.

That was most evidenced in 2010, when Sebastian Vettel triumphed in a four-way showdown against Fernando Alonso, Mark Webber and Lewis Hamilton, and in 2014 and 2016 season-long battles between Mercedes teammates Lewis Hamilton and Nico Rosberg were settled at the last, Hamilton triumphing in 2014 and Rosberg in 2016. Abu Dhabi's most infamous title decider took place in 2021 when a titanic tussle between Hamilton and Max Verstappen was settled on the final lap in acrimonious circumstances, amid the dubious application of regulations concerning the race restart. Verstappen, on fresh tyres, prised the world title away from Hamilton after a wheel-to-wheel battle, but the controversy from the fallout crackled through the paddock and will always be contentious.

If matters have long since been settled, then there is a definitive end-of-school vibe, though the recent expansion of the schedule – with the last race now just a few weeks away from Christmas – has added to the aura of *Let's get this done and head back home*. When Abu Dhabi joined the calendar in 2009, it was the last of 17 rounds, but it is now the final of 24, meaning colds, coughs and flu hurtle around personnel, gradually worn down by long-haul flights, long working hours and the drying impact of flitting between various air-conditioned spaces and hotel rooms. The 'paddock pandemic', if you will.

Those who have enjoyed a successful season are eager for one last blast in a competitive car and the prospect of further riches. Those who have had a dispiriting campaign find every day a drag and cannot wait to climb out of a car they have hated for the final time.

The only good thing about that race is it was the last race, has been said by more than one driver after an Abu Dhabi Grand Prix.

Those at the sharp end of the grid will caution that next year might be different and those further down the field like to emphasise that lessons have been learned for the following season. Reviewing the season can bring smiles and frowns, memories of highs and lows, along with regrets and missed chances.

Yet even in a 'dead rubber' – a race when a title has already been settled – there is still often something to play for: most notably championship positions further down the pecking order. Drivers outside the front-running places will care little for their own classification, even if there is pride at stake, and even for hardcore Formula 1 fans such scraps are hardly pulsating. But for the teams each position is vitally important. Each position in the championship is worth millions in prize money – with the difference estimated around $10 million per position depending on Formula 1's overall revenues – and some team members' bonuses will be riding on where the outfit finishes. There is also, of course, the simple matter of competitive pride.

Formula 1 visits some awe-inspiring circuits across the course of its nine-month voyage, but the season finale track is not one that gets the juices flowing. It is merely fine, which is a tad underwhelming for the final race of the season when the likes of Suzuka and Interlagos are available – but money talks and Yas Marina pays a premium to host the finale. There are very few challenging turns, it is plentiful of run-off and it is not a circuit renowned for its thrilling competition, despite some partial revisions in 2021. A sequence of off-camber corners at the Turn 6/7 chicane and in the slower final sector also frustrates drivers.

'They changed the track layout a couple of years ago and generally, although it's made the driving experience a little better than before, I don't feel it has affected overtaking opportunities, so it hasn't improved the actual racing,' said Daniel Ricciardo.

The event nevertheless still has its slight quirks. For its 2009 debut Yas Marina became Formula 1's first twilight race, and lights out at 17:00 means the grand prix begins in daylight and ends at night. Teams and drivers must be careful not to head down the wrong path in the first and third practice sessions, which take place in the heat of the day, unlike second practice, qualifying and the race itself. The Yas Marina Circuit is also traversed by the impressive and eye-catching 499-room W Hotel. The two buildings, perpendicular to each other, are connected by a couple of walkways and covered by a honeycombed lattice. They feature 5000 diamond-shaped glass panels, stacked with LEDs, ensuring the W Hotel can put on a lights show for special events and is illuminated during most of the track activity. The marina behind the paddock, which perches alongside parts of the final sector, adds to a chilled atmosphere, though it's not quite Monaco in size, energy or glamour.

The end of the season means the closing of a year-long chapter. There can be drivers leaving their teams, team members departing to wear new colours or take in a different working environment, or other members of the paddock swiping in and out for the final time. Those moving teams – including drivers – often have some form of forfeit, which can conclude with them ending up in the marina located directly behind the hospitality buildings. One year one team member lost a wedding ring after being dunked into the marina. Diving equipment was sourced, along with an underwater metal detector, but after an hour the valiant effort proved fruitless. A saviour appeared, in the form of a local diver from whom the detector was borrowed, and the following day he unearthed the ring after a two-hour search. Drivers in the past have been duct-taped to trolleys (and doused in god knows what) before being delivered to their new team. It can also be a moment for sad farewells and more sombre reflections: if a driver of high esteem is walking away into retirement, paddock personnel pack on to the assigned rooftop of a hospitality unit for a gathering.

And when the chequered flag falls on Sunday it brings down the curtain on the campaign and the night sky above Yas Marina is illuminated by an ostentatious fireworks display. The winner – and champion – often performs doughnuts on the straight before the celebrations begin, despite the best efforts of engineers to ask them not to damage components. Several other drivers often join in, just for the hell of it.

Everyone has shared a journey through the roller coaster of a season: blood, sweat and tears, a range of emotions, and highs and lows. People drift off out of the paddock, some to zone out, others to party the night away to celebrate the end of the season.

That's that.

And yet it also isn't.

Come Monday morning, a swathe of personnel will filter back into the paddock to prepare the cars for a day of post-season testing on the Tuesday.

The one-day test is arranged twofold. One car is allocated to regular race and test drivers in order to carry out testing on behalf of tyre supplier Pirelli, facilitating the Italian company in gathering additional data, particularly when it is assessing new compounds or constructions for the following season. The second car is driven by a rookie driver – classified as someone with two or fewer grand prix starts – in accordance with a rule that was introduced in order to enable the next generation to gain understanding of

Formula 1 machinery and how teams operate. This is often a first opportunity to see a rising star behind the wheel of a Formula 1 car, who is able to drive in a low-pressure environment, away from the heightened scrutiny of a grand prix weekend. Stable warm weather, known track conditions and a gripped-up track after a weekend of Formula 1 activity help the young drivers. It can also be a chance for teams to put some new mechanics and engineers into the working environment of a Formula 1 garage and pit lane for the first time, building up their experience.

On occasion there is fervent interest in the one-day test if a driver is moving teams for the following season. Contracts usually run for a 12-month calendar period, but regularly agreements are reached – particularly if multiple parties are involved – for a driver to begin work with their new team for the first time. This usually involves a driver wearing a plain race helmet and a sponsor-less suit, and can extend to a stripped-down car livery, due to the complexities of clashing sponsorship deals. That creates a flurry of activity throughout the day: there is a natural interest in seeing a driver in new surroundings and the activity gives both parties early pointers on what is needed ahead of the following season.

The paddock during the post-season test resembles an external warehouse, with forklift trucks scurrying around carrying hundreds of boxes and pallets ready return to European factories – or even to be shipped out to the opening event of 2024.

Teams will return to European bases and continue working on the following year's cars, marketing and communication teams will already be busy scheduling activities for car launches and the early season grands prix, while drivers will have some simulator and marketing duties before being allowed to leave for holidays.

There is also the matter of end-of-season parties and gatherings, most notably the FIA Gala, which typically takes place around a week after the final grand prix of the season. The location for these vary: Paris is the central home of the FIA, but they have in the past taken the Gala to Bologna, St. Petersburg, Vienna and Baku. And while the event is a celebration of all FIA-sanctioned motor sport, Formula 1 is often the focal point and the culmination of the ceremony – and it is mandatory for the top three in the drivers' championship to attend, in order to collect their respective trophies. The world championship trophy, which resembles a large screw, features all the names and signatures of each champion dating back to 1950. At the end of a lengthy and sometimes bewildering ceremony, the drivers finishing second and third in the standings no doubt ponder whether finishing fourth would have been a preferable outcome to their season.

There is a chance to celebrate, commemorate or commiserate about the season. The teams, after all, are extended families, who spend nine months of the year on the road together. But the focus is already on the following season, the start of which looms on the horizon . . .

The stunning Sheikh Zayed Grand Mosque is one of Abu Dhabi's landmarks

EPILOGUE

By the time the final round of the season rolls around there is always someone who pipes up. *You know*, they start, *it's only 95 days until practice for the first race.*

Usually such a quip is met with a glare: by the end of the campaign most people are ready to zone out from racing for a few weeks, reflect on the season – and enjoy the fast-approaching Christmas and New Year period – before even thinking about the following year. Though, behind the scenes, the next season will have been on the drawing board almost since the previous campaign ended 12 months beforehand.

There are tens of thousands of people who make the sport go around. Beyond the 2000+ on the ground at each event – the circuit personnel, volunteer marshals, logistics teams, medical crews, all swiping in and out of the turnstiles each day – are hundreds and hundreds more at each factory, some of whom will be undertaking gruelling night shifts; those operating in the supply chain; and the marketing and communication armies.

Each Formula 1 team is a highly sophisticated and professional operation that continues to be refined year on year, with every member being of the utmost importance; think of it as the various sections of an orchestra, all conducted by the Team Principal.

The camaraderie inside teams has to be on a level that is above and beyond the usual office environment. This is a landscape in which each member of the team is responsible for and dependent upon everyone else. People travel the world together, have several meals a day together, socialise together, share long journeys around the world together, and it is a lifestyle that demands full commitment, representing a high-class brand in pursuit of perfection. That can be draining, especially during big time shifts or high-pressured environments, and there are sacrifices to be made: missed birthdays, anniversaries, other milestones, and a challenge to balance Formula 1 life and a normal life.

This camaraderie extends to the relationship between different teams, who are competitors and rivals on-track, but part of a dysfunctional family off-track, stepping in to help out in times of crisis or need. Friendships and bonds are maintained even when the team gear is different: everyone respects everyone else, irrespective of which particular pit garage or hospitality building you call home.

These homes are never permanent, but are ever-changing, sometimes on different continents on successive weekends, as Formula 1 continues to expand its reach across the globe.

There are the luscious valleys of Austria, where different shades of green dominate the horizon as trees blanket the mountain sides, where the dawn chorus breaks at 4 a.m. in the height of summer, and where it feels as if the environment has not significantly evolved in decades, and is never likely to; or the dense streets of the ever-expanding metropolis Singapore, where cars pound the city roads long after the sun has set and everyone sweats buckets. There are venues that may be past their best, flaking and decaying, or venues with a century of stories, ghosts and folklore. Finally, there are some venues where the paint is still drying and where only the opening couple of chapters have been written. The best of these events would be nothing without the fans, who flock through the gates come what may, embracing the action from dawn until dusk, whether they have been hardcore enthusiasts for decades, or newer devotees keen to learn more, taking the chance of fuelling their fresh passion at close quarters.

The technological race is relentless. A component crafted, developed and honed will be superseded and redundant just a few months on from its introduction after the announcement of the next specification . A car that the world is eager to see in February, and which carries the hopes of a team for a season, will become an exhibit in December, its purpose complete as soon as the chequered flag falls at the final race of the year. A machine once deemed futuristic and groundbreaking, packed with the latest aerodynamic devices and gizmos, will eventually find itself propped up in a museum, or wheeled out for historic demonstration runs, overtaken by the next generation of cars. Cars yet to be dreamed up will swiftly be deemed old-fashioned and there will be a nostalgia for our present. Some of the greatest minds in science and engineering come up with fresh ideas, new approaches and left-field solutions, in a bid to outdo the opposition. It is a persistently fascinating game, as technical chiefs take a look at rival machines and hope that their latest batch of goodies will yield the hoped-for lap time.

One year will roll into the next, a relentless cascade of events, commitments, heroes and villains, and new chapters being written into the annals of Formula 1.

So where to next and what will the future bring?

The composition of the grid will gradually evolve. Race winners and champions will eventually hang up their helmets, some raging against the decline of their own powers, to be replaced by a fresh crop of enthusiastic youngsters eager to walk in their footsteps, to undertake the same learning process and to inscribe their names on the trophies amassed by the greats of the sport. The current class will be idols for the next generation, who are now contesting karting and on the pathway to joining Formula 1 in the 2030s and beyond, with both genders hopefully represented. They will one day be the record-breakers and the heroes, reminiscing about getting home from school and watching the likes of Max Verstappen, Lando Norris and Charles Leclerc.

Formula 1 is currently enjoying its longest period of stability in terms of team composition: owners Liberty Media are happy with the 10-team structure and unwilling to allow further entries for at least the next few years. Each entry is now worth over £1 billion and there are financial ramifications if the revenue pot is going to be split with an 11th or even 12th team, given that Formula 1 cars are in effect fast-moving billboards. This is an understandable approach, but also a shame to some degree, considering the size of grids in other categories, because 12 healthy and financially stable teams – operating within a workably sustainable structure – would be a boon for Formula 1. Not merely for competition's sake, but also for greater opportunities for drivers and personnel. Many top drivers, mechanics, engineers, strategists and so on got their foot in the door somewhere, before rising up the chain as the years wear on.

Hopefully the culture will change and new teams will be allowed before long – after all, the likes of Red Bull Racing and Aston Martin grew out of junior teams who fancied a crack at Formula 1. On a political level the cordiality of the relationship between the teams, Formula 1 and the FIA (its governing body), ebbs and flows. The *divide and conquer* tactics utilised by past regimes has gradually been replaced with a more holistic approach, as Formula 1 learns and adapts from other sports. Even so, it does not take long for any simmering tension or feelings of angst to rise to the surface, amid the never-ending quest for power and control.

Formula 1's calendar is now the longest it has ever been; only January and February – when the midnight oil is being burnt back at the factories – have no grands prix and at 24 events the schedule already risks reaching saturation point. Some argue that such a stage was reached when the calendar exceeded 20 grands prix. As compelling and enticing as Formula 1 is, there remains the danger that competitors and highly skilled personnel have shorter careers due to the burden of the schedule: more time away from home, more time in identikit hotel rooms, more time in soulless airport terminals, more time staring out of aeroplane windows, or worse, hemmed in the middle seat in economy. Only the top level of the paddock have the luxury of privately chartered jets or business class and the travelling can develop into a grind.

How will the calendar evolve? It is highly unlikely to get shorter, given the financial injection to Formula 1 and its teams that arrives courtesy of a grand prix, and there is the possibility that a 25th event could be added, or the existing grand prix weekend format modified to make each day a compelling proposition. The biggest alteration to the weekend schedule came in 2021, when Formula 1 introduced a Sprint race at three grands prix, which was expanded to feature at six rounds from 2023. On Sprint weekends there is only a single practice session, rather than three, with an additional qualifying session and a Sprint Race – which is around a third distance of the grand prix and lasts for approximately half an hour. It means that while there are 24 grands prix there are now 30 'races' per year, and the popularity of these Sprint events remains hard to gauge. Some drivers enjoy them, but others are warier – Max Verstappen is among the most vocal critics. Formula 1 has been swift to highlight the stronger TV numbers and social media engagement during Sprint events, as the *sport versus entertainment* argument rumbles on. Statistics can always be interpreted and skewed depending on the demands.

Several events are stronger than they have ever been, with record crowd numbers year on year, and a noticeably younger and broader demographic. Airport terminals pre- and post-race weekends are, for the most part, now festooned with fans bedecked in merchandise from a range of teams and drivers, the kind of scenario that happened only occasionally in past decades, particularly in the United States. Formula 1 is increasingly mainstream.

There will be new grands prix – Madrid will host the Spanish event in 2026 – and it is inevitable that some will fall by the wayside, due to calendar space, a lack of available finance – perhaps caused by the loss of a backer or influential figure, or perhaps by a change in the country's circumstances. Italy holding two grands prix is thought unsustainable long term, while Belgium and the Netherlands remain at risk, particularly once Verstappen's career comes to a close. Several venues at least have long-term contracts to be able to plan, investing in upgraded infrastructure and facilities to ensure they can keep pace with Formula 1's increased demands. There has always been a desire to blend historic circuits with new venues, respecting the past, never resting on any laurels.

Formula 1 is happy with three grands prix established in the United States and is keen in the short to medium term to solidify what it has, particularly given the relative newness of Miami and Las Vegas.

Accelerating a footprint in China, and getting a stronger stranglehold in Far Asia, is also an ambition. Korea is heavily mooted for a comeback – although not at the briefly used and unloved circuit in Mokpo – while Thailand's administration is keen to bring Formula 1 to the country for the first time and to show off the streets of capital Bangkok.

Formula 1 is eager to hold a race in Africa, which would give it a presence on each habitable continent, but it has not visited since 1993. And the front runner, South Africa, remains beset by issues that have so far presented only insurmountable barriers. A possible return to the permanent Kyalami circuit is for now on the back burner.

The ongoing question is: *How much more can Formula 1 continue to grow – and what does that growth mean?* Some grands prix are at physical capacity and cannot expand unless ticket prices are hiked, which has already sadly happened at some locations, pricing out spectators. Liberty Media has enjoyed a period of stunning growth as it nears the end of its first decade of ownership, but shareholders demand progress, so what next? Where will Formula 1 evolve in the increasingly congested global sporting landscape?

There will be new regulations as Formula 1 reacts to the global landscape, whether that be the pure sporting spectacle, the financial situation or the technological scene. Formula 1 strives to remain road relevant and future-thinking, with the next batch of technical regulations in 2026 set to welcome updated chassis, revised power units and fully sustainable fuels. That has enticed Audi to Formula 1 for the first time in the history of the German automotive giant, which will form a strategic works partnership with Sauber Motorsport. The ambition for the sport is always to create racing that is close and exciting without being artificial, while evolving with the world that surrounds Formula 1 – a tough task, and one that is always likely to be unrealistic. Even the sport's officials, like the teams that they regulate, are pursuing an improbable perfection, tasked with sculpting the future while respecting what came before.

What does Formula 1 itself want to be? And what should it be?

Formula 1 has never been static and has always evolved, whether that be the calendar, the regulations or the protagonists. The Formula 1 of the 2020s is not the Formula 1 of the 2000s, and while there is a common thread the championship will continue to transform as it creeps towards its centenary.

But fundamentally the ethos remains the same. Formula 1 remains the pinnacle of four-wheel motor sport, with the fastest cars, the most breathtaking locations and some of the finest drivers in the world. Their goal remains the same as those who first started out in motor sport: to win, to become champion and to have fun during the process. And just as in the 1950s spectators flocked to stand adjacent to the roads tackled by the drivers, so too in the 2020s do fans travel to circuits to watch masters at work behind the wheel of ever more complicated machinery and to say, *I was there*.

Each season is enthralling, intoxicating, demoralising, entertaining, tedious, emotional, addictive and draining, sometimes all at the same time, and phrases such as *See you in Singapore* or *When are you getting to Mexico?* become as normal as *See you tomorrow*. It isn't perfect – no sport, job or lifestyle is – but it is a beguiling championship full of characters, which will continue to motor on year on year.

Formula 1 remains the pinnacle of motor sport, relentlessly evolving, constantly developing.

ACKNOWLEDGEMENTS

To Matt and the team at Bloomsbury Sport for taking a manuscript and shaping it into something readable, and for their patience and effort in getting everything over the line.

To Melanie, my agent, for taking a chance and for her persistence.

To everyone interviewed and featured in this book for providing their time and insight: Alan Adler, Alex Jacques, Andrew Westacott, Ariane Frank-Meulenbelt, Beat Zehnder, Bobby Epstein, Bruno Michel, Francois Dumontier, Christian Menath, Daniel Ricciardo, Davide Valsecchi, Dean Locke, Erik van Haren, Florent Gooden, Frankie Mao, Jody Egginton, Jon Malvern, Jesus Balseiro, Lando Norris, Lawrence Barretto, Steve Hill, Marco Perrone, Mark Norris, Oscar Piastri, Pete Crolla, Rodrigo Sanchez, Rupert Manwaring, Sandor Meszaros, Sheikh Salman bin Isa Al Khalifa, Stoffel Vandoorne, Tyler Epp, Tom Wood, Zhou Guanyu, Valtteri Bottas and Yuki Tsunoda.

To the people of the Formula 1 paddock for keeping the show on the road, providing entertainment to millions of people across the world.

To the communication teams at Formula 1, the FIA, at each of the 10 Formula 1 teams, and at circuits – in particular Liam Parker, David Leslie, Tom Wood, Roman de Lauw, Tom Cooney, Harry Bull, Will Ponissi, Fabiana Valenti, Stuart Morrison, Jess Borrell, Carla Corbet, Laurence Jones, Alexa Quintin and Sandrine Garneau.

To the Groovers – you know who you are (for better, for worse) and this job would be significantly less fun if it wasn't for you lot – and to Kate and Ju, for somehow adopting me and keeping me sane.

To Yorkshire Tea and McVitie's Digestives, for their mere existence.

To my parents, for their unwavering support and love.

INDEX

60th anniversary, F1 57
1000th grand prix, China 33–4

Abu Dhabi country and culture 240–1
Abu Dhabi Grand Prix: Yas Marina
  Circuit 60, 240–3
academies, F1 team 33, 158–60
accommodation, team 94–6, 186–7
Adler, Alan 213, 218
age and training regimes 4–5
AI-Ring, Austria 126
airflow and design concepts 8
Al-Hamad, Amro 233
Albers, Christijan 167
Albert II of Monaco, Prince 102
Albon, Alex 112, 138
Alboreto, Michele 175
Alfa Romeo 135, 144
  see also Sauber Motorsport
Alonso, Fernando 32, 49, 52, 58, 92,
  100–1, 110–11, 114, 155, 218, 221, 241
AlphaTauri 92, 125, 177
see also RB, Visa Cash App
Alphen, Henri van 166
Alpine 33, 137, 159, 160
altitude issues, Mexico City 210–12
Andretti, Mario 46, 202
Arab Spring (2011) 59
Aramco oil 63
Argentinian Grand Prix 221
Arnoux, René 151
Ascari, Alberto 144, 179
Ascari, Antonio 144
Aston Martin 63, 114, 116, 137, 149,
  195, 250

Atack-Martin, Faith 194
Audi 252
Austin, Texas city and culture 202–3
  see also United States Grand Prix:
    Circuit of the Americas, Austin
Australian Grand Prix: Albert Park
  Circuit, Melbourne 15–25, 57, 73,
  120, 129, 185
Australian Grand Prix Corporation
  (AGPC) 21–2, 24
Australian Grand Prix in Adelaide 19, 24
Austrian Grand Prix: Red Bull Ring,
  Spielberg 124–30, 168, 248
Auto Union 156
Automobile Club de Monaco 98
Automobile Club d'Italia 179–80
Azerbaijan country and culture 181–2
Azerbaijan Grand Prix: Baku City
  Circuit 120, 181–3, 188

Bahrain country and culture 54–7, 59
Bahrain Grand Prix: Bahrain International
  Circuit, Sakhir 16, 42, 54–61, 74,
  208, 232, 239
Balseiro, Jesús 108, 109, 110–11, 112,
  113–14
Bandini, Lorenzo 212
Barretto, Lawrence 40
Barrichello, Rubens 217
Baumgartner, Zsolt 156
Bearman, Oliver 133
Belgian Grand Prix (1985) 20
Belgian Grand Prix: Circuit de
  Spa-Francorchamps, Stavelot 41,
  143–7, 149–51, 168

Berger, Gerhard 89, 125, 175
Bianchi, Jules 50
Bin Salman, Mohammed 62
Bolsonaro, Jair 214
Bondurant, Bob 144
Bottas, Valtteri 1, 13–14, 19, 73, 74, 164, 185, 186–7, 210
Boyan, Todd 79, 80
Brabham, Sir Jack 16
Brands Hatch circuit, UK 134
Bratches, Sean 77
Brawn Grand Prix 45
Brazilian Grand Prix 214, 215–16
Bridgestone tyres 235
Bristow, Chris 144
British Grand Prix, Silverstone 41, 106, 133–42
   the track 135–6, 140–1
broadcasting production/arrangements, F1 TV 12, 35–43, 129, 132
Brown, Zak 96
budget for cars 8
Button, Jenson 45, 58, 133, 155, 219

calendar/schedule evolution, potential 250–2
Campos, Adrián 110
Canadian Grand Prix, Circuit Gilles Villeneuve, Montreal 115–20, 123, 129
Carey, Chase 224
cars
   chassis shape 8, 9
   design 6–10, 248–9
   launch 10–12
   testing 12–14, 244–5
celebrity fans 41, 81
China Central Television (CCTV) 35
China, F1 market 33–6, 251
Chinese Grand Prix: Shanghai International Circuit 31–6
Christmas and New Year shutdown 8
circuit accommodation, team 93–6
Clark, Jim 133, 138, 212
Colombian Grand Prix proposal 221
commentators, F1 world feed 39–40
component production 9–10
contracts market, F1 165, 245
Correa, Juan Manuel 146

COVID-19 global pandemic 20–1, 32, 33, 37, 46, 77, 90, 128–9, 131, 141, 169, 177–8, 180, 232–3
crash tests 8–9
crashes 42, 49–51, 58–9, 63–4, 89–90 112–113, 144, 146–7, 149, 221
   TV coverage protocols 41–3
Crolla, Pete 76, 82, 83–4, 85, 86, 163

Dakar Rally 67
Damiani, Angelo Sticchi 180
de Crawhez, Baron Joseph 143
de la Rosa, Pedro 110
de Portago, Alfonso 110
design, car
   see cars
DHL 82
disqualifications 17, 75
Domenicali, Stefano 91, 142, 151, 221
Donington Park circuit, UK 134
Doornbos, Robert 167
Drapeau, Jean 116
Drive to Survive Netflix documentary 34, 199–200, 224
drivers' briefing, Grand Prix 75
Dumontier, François 115, 116, 118, 119, 123
Dutch Grand Prix: Circuit Zandvoort, Zandvoort 166–72

Ecclestone, Bernie 54, 55, 56, 153, 200, 214
Egginton, Jody 8–9, 10, 12, 25
Eifel Grand Prix (2020) 131
Emilia Romagna Grand Prix: Autodromo Enzo E Dino Ferrari, Imola 88–93, 180
end-of-season in Abu Dhabi 243–5
engine manufacture, Japanese 45
Epp, Tyler 80, 81
Epstein, Bobby 199, 200, 204
European Grand Prix 109, 110, 182
Event Technical Centre, onsite F1 36, 37
Extreme H (formerly Extreme E) 67

F1 Academy 160–1
F1 Digital+ 38
Faenza, Italy 92–3
Fangio, Juan Manuel 221

INDEX

Far Asian potential locations 251–2
Farina, Nino 135
fatalities 50–1, 89–90, 109, 110, 116, 124, 144, 146–7, 205–6, 216
Fédération Internationale de l'Automobile (FIA) 7, 8, 31, 42, 69, 70, 75, 94, 115, 116, 120–2, 128, 163, 232, 237, 245, 250
    Contract Recognition Board 160, 165
    end of season Gala 245–6
    Safety Department 147–9
    see also rules and regulations
Ferrari 8, 11, 33, 60, 75, 91, 92, 102, 110, 112, 132, 159, 173–6, 178, 179, 200, 205, 212, 217, 230
    Tifosi/fans 173, 174–5, 178, 179
Ferrari, Enzo 174, 175
financial/elitist boundaries, F1 158–9, 160
Fisichella, Giancarlo 221
fitness check-ups 4
Fittipaldi, Emerson 214
FITV 38, 39
food/diet 187, 194, 210
Formula 2 and 3 17, 21, 33, 70, 157–8, 159, 167, 189
Formula 4, women's 160–1
Formula E 67
Formula Motorsport 157, 161–2
Frank-Meulenbelt, Ariane 153, 155, 156
French Grand Prix 151–2, 156
fuel 69–70
Fuji Speedway, Japan 45–6
Fuji Television 37

gaming rigs/simulation games 6, 228
Garde, Giedo van der 167
Garfinkel, Tom 78
Gasly, Pierre 146, 164, 168, 177–8, 181, 187, 215, 232, 234
George VI of Great Britain, King 135
German Grand Prix 131–2
Gibson, James 226
go-karting 6, 16, 18, 33, 68, 142, 157, 167, 228
González, Federico 207–8
Gooden, Florent 105, 106, 107
grid composition and size, evolving 249–50
'grid kids' 18
Grosjean, Romain 42, 58–9

Haas 52, 58–9, 73, 82, 117, 137, 163, 194, 202, 220
Häkkinen, Mika 46
Hamad Bin Isa Al Kalifa, King 57
Hamilton, Lewis 32, 47, 52, 58, 64, 75, 90, 110, 112–13, 118, 133, 138–9, 141–2, 167, 175–6, 177, 183, 197, 198, 199, 201–2, 206, 212, 215, 217, 219, 221, 235, 241–2
Hard Rock Stadium campus, Miami 77–82
Haren, Erik van 168, 171
Hawthorn, Mike 133
heat and hydration management 3, 191–5, 210, 236–7
Hill, Damon 19, 133, 134, 149
Hill, Graham 133, 144, 212
Hill, Steve 224
Hitech 137
Hockenheim circuit, Germany 131, 132
homologation tests 8–9
Honda 44–5, 53, 159
Honda, Sochiro 45
Horner, Christian 127–8, 169
Hubert, Anthoine 146–7
Hülkenberg, Nico 124, 126–7, 133, 135, 220
Hungarian Grand Prix, Hungaroring, Budapest 45, 153–7
Hunt, James 46, 133

IFEMA exhibition centre circuit, Madrid 113–14, 251
Igora Drive circuit, Saint Petersburg 184
Île Notre-Dame circuit, Montreal 116
Indianapolis Motor Speedway circuit 198, 200
interviews, driver 40–1
Italian Grand Prix: Autodromo Nazionale Monza, Milan 88–9, 92, 98, 124, 153–4, 173–80

Jacarepaguá Circuit, Brazil 214
Jacques, Alex 39–40, 42–3, 141, 179
Jaguar Racing 125
Japan country and culture 44, 46, 47–9, 51–3
Japanese Grand Prix, Fuji Speedway (1976) 45–6

Japanese Grand Prix: Suzuka International Racing Course   44–53, 129
Jarama circuit, Madrid   108–9
Jerez circuit, Spain   109
jet lag and time adjustment   184–7, 195–6, 229
Jones, Alan   16
Jordan   149
Jordan Grand Prix   137

Key, James   26
Kobayashi, Kamui   52
Komatsu, Ayao   52
Kubica, Robert   118
Kvyat, Daniil   18

La Meuse Cup   143
Las Vegas city and culture   224–5
Las Vegas Grand Prix: Las Vegas Strip Circuit   98
Lauda, Niki   46, 124–5, 126
launch of new cars   10
Leclerc, Charles   49, 50, 62, 64, 75, 88, 91, 101–3, 146–7, 159, 175, 223, 228, 229, 230
Liberty Media   33, 77, 169, 190, 199–200, 212, 225, 227, 231, 249–50, 252
Locke, Dean   36, 38, 39, 41, 42
logistics, global travel/cargo   82–6
London Grand Prix proposal   141–2

Ma Qinghua   32–3
McCullough, Tom   195
McLaren, Bruce   197
McLaren   10–11, 13, 17, 19, 45, 60, 73, 89, 95–6, 117, 137, 139, 145, 159, 160, 175, 180, 197, 216
Maffei, Greg   225
Magnussen, Kevin   117, 166, 170, 184, 191–2, 203, 220
Magny-Cours circuit, France   152
Maldonado, Pastor   112
Malvern, Jon   3–4, 186, 187, 192, 193, 210–11
Mansell, Nigel   19, 133, 138
Manwaring, Rupert   185, 186, 195
Mao, Frankie   31, 33–5
Martínez-Almeida, José   113
Massa, Felipe   32, 217
Mateschitz, Dietrich   125, 126, 127–8

Media and Technology Centre, F1 Biggin Hill   36, 37
medical facilities and cars   148
Melbourne city and culture, Australia   15–16, 19, 22–4, 57
Menath, Christian   131, 132
Mercedes   32, 45, 73, 74, 75, 124–5, 128, 136–7, 156, 159, 167, 175, 200
merchandise, Grand Prix   52, 171, 230
Messi, Lionel   233
Mészáros, Sándor   154, 156–7
Mexico City city and culture   206–8
Mexico City Grand Prix: Autódromo Hermanos Rodríguez, Mexico City   205–12
Miami Grand Prix: Miami International Autodrome   41, 76–82, 197, 224
Michel, Bruno   157, 158
Michelin tyres   198, 235
Minardi   92, 110, 125
Monaco Grand Prix: Circuit de Monaco, Monte Carlo   ix–xii, 37, 73–4, 94–105, 106, 107, 229, 230
Monaco principality and culture   103–4
Montjuïc Park, Barcelona, Spain   108–9
Mont-Tremblant circuit, Quebec   115
Monza *see* Italian Grand Prix: Autodromo Nazionale Monza, Milan
Moss, Stirling   144, 197–8
motorhomes, team   94–6, 186
Motorsport Park, Ontario   115

National Football League (NFL), US   78, 79, 81
neck strength and endurance exercises   3
New York Grand Prix proposal   76–7
Nguyen Duc Chung   190
Nivelles circuit, Belgium   145
Norris, Lando   3, 50, 81, 100, 133, 139–40, 159, 237
Norris, Mark   95, 96, 180
Nürburgring, Germany   131

Ocon, Esteban   146, 152, 153, 155–6, 191
opening campaigns, Formula 1   24–30, 57, 61
Ophem, Henri Langlois van   143
Österreichring, Austria   125–6

Pace, Carlos 216
partnerships/sponsorships 12, 34–5, 233
Paul Ricard circuit, France 152
Peréz-Sala, Luis 110
Pérez, Sergio 64, 80, 183, 202, 208–9
Perrone, Marco 93, 163
Pescara Grand Prix 180
photographers, trackside 105–7, 121, 129
Piastri, Oscar 13, 15, 17–18, 22–3, 138, 160, 241
Piquet, Nelson 89, 214
Pirelli tyres 74, 94, 235–6, 244
Pironi, Didier 151–2
pit stops 41, 71–3
Ponissi, Will 11, 122
pre-season training 1–6
press media/journalists 120–3, 129
production team, F1 world feed coverage 36–42
Prost, Alain 46, 134, 151, 175, 198
Provera, Marco Tronchetti 236

Qatar country and culture 237–8
Qatar Grand Prix: Lusail International Circuit 73, 232–9
Qatar Motor and Motorcycle Federation 233
qualifying sessions 25–6, 40, 69, 89

Räikkönen, Kimi 219, 221
Ratzenberger, Roland 89
RB, Visa Cash App 92, 94, 125, 137, 159, 167
Red Bull 125, 126, 127–8, 129, 136, 159
Red Bull Racing 8, 32, 45, 73–4, 113, 125, 127–8, 129–30, 136, 137, 138, 159, 167, 168, 169, 208, 250
refuelling, mid-race 69–70
regulations *see* rules and regulations
Reid, Kate 15–16
Renault 110, 117, 137, 152
Ricciardo, Daniel 16–17, 18, 26, 65, 74, 93, 97, 100, 118, 125, 182, 202, 205, 209, 212, 218, 221, 230, 240, 243
Rindt, Jochen 124, 126
Riverside circuit, California 197–8

rock/pop concerts 182, 201, 209–10
Rodríguez, Ricardo and Pedro 205–6, 208
Rosberg, Nico 32, 58, 112–13, 131, 182, 206, 241
Rossi, Alexander 202
Royal Automobile Club of Belgium 144
Royal Automobile Club of Great Britain 134–5, 138
rules and regulations 221, 237, 242
  driver incidents and Race Control 75
  fuel regulations 69–70
  post-race checks 75
  pre-raceday car seals 69
  safety 147–9
  technical regulations 6–7
  testing regulations 12
Russell, George 44, 49, 74, 133, 149, 159, 163, 228
Russian Grand Prix 183–4

Safety Department, FIA 147–9
Sainz, Carlos 91, 111, 168, 193, 219, 230
Salman Al-Khalifa, Sheikh 54, 55, 56, 57, 59–60
Sanchez, Rodrigo 206, 207, 208, 209–10
São Paulo city and culture 219–20
São Paulo Grand Prix: Autódromo Jose Carlos Pace 120, 213–22
Sargeant, Logan 5, 202, 228
Sato, Takuma 52
Sauber Motorsport 8, 11, 26, 33, 122, 125, 159, 164, 252
Saudi Arabia country and culture 16, 54–5, 62–3, 65–7, 239
Saudi Arabia Grand Prix: Jeddah Corniche Circuit 60, 62–8
Saudi Motorsport 67–8
Schlesser, Jean-Louis 175
Schumacher, Michael 19, 32, 46, 58, 64, 90, 109, 112, 131, 134, 150, 174, 175
Second World War 134–5, 144, 166
Senna, Ayrton 46, 89–90, 134, 175, 198, 214, 215–16
Sepang circuit, Malaysia 56, 190
Shanghai city and culture, China 31–2

Sheinbaum, Claudia 206
Silverstone *see* British Grand Prix, Silverstone
simulator training 6, 228
Singapore Grand Prix: Marina Bay Street Circuit 98, 106, 129, 189–96, 248
Smeets, Sven 159
social media 11, 35, 38
South Africa Grand Prix potential 252
Spanish Grand Prix: Circuit De Barcelona-Catalunya 13, 108–14, 168
Sprint races, F1 250–1
Stacey, Alan 144
Stewart, Sir Jackie 55, 133, 144, 149
Stommelen, Rolf 109
Stroll, Lance 116–17, 137
Styrian Grand Prix (2020) 128–9
summer break, F1 162–5
Surtees, John 133, 212
sustainability considerations 46, 70, 118, 252
Suzuki, Aguri 52
Sydney, Australia 19–20, 24
Szisz, Ferenc 156

Tambay, Patrick 151–2
Taylor, Mike 144
team camaraderie 247–8
team headquarters/facilities in the UK 136–8
team structure and strategy, race-day preparation 68–73
Télé Monte Carlo 37
television coverage *see* broadcasting production/arrangements, F1 TV
test drives 6, 244–5
testing 12–14
Tilke, Carsten 63, 227–8
Tilke, Hermann 56
Todt, Jean 175
Toro Rosso 92, 125, 167, 177
  see also RB, Visa Cash App
Toyota 46
training regimes, physical 2–4
travel logistics, global 82–6
  see also jet lag and light adjustment
Tsunoda, Yuki 5, 48, 52, 92–3, 138
Turki Al Saud, Prince Abdulaziz bin 65–6

Turkish Grand Prix 183
Tuscan Grand Prix 180

United Arab Emirates (UAE) 239, 240–1
United States Grand Prix: Circuit of the Americas, Austin 75, 197–204, 206, 224
United States Grand Prix, Phoenix 198

Valencia circuit, Spain 110, 217
Valsecchi, Davide 173, 177, 179
Vandoorne, Stoffel 143, 145, 151
Verstappen, Jos 167, 168
Verstappen, Max 6, 46, 64, 68, 81, 90, 107, 113, 115, 116, 125, 129–30, 141, 150, 167–8, 169, 171–2, 189, 193, 199, 221, 231, 235, 241–2, 251
  Orange Army/Dutch Fans 130, 155, 168, 171
Vettel, Sebastian 32, 46, 58, 89–90, 110, 125, 131, 132, 174, 177, 199, 241
Vietnam Grand Prix proposal 190
Villeneuve, Gilles 116
Villeneuve, Jacques 19, 109, 116

W Series 161
Watkins Glen circuit, New York 198
Watkins, Professor Sid 149
weather issues 47, 118, 149, 191–5, 199, 211–12, 220–1, 229–30, 235, 236–7
Webber, Mark 16, 18, 92, 221, 241
weight maintenance, body 3
Westacott, Andrew 21, 23, 24
Whitaker, Martin 67–8
Williams 10, 19, 45, 112, 137, 159, 202
winter break 1–6
Wolff, Susie 161
women drivers 160–1
Wood, Tom 120–1

Zehnder, Beat 162, 164
Zhou Guanyu 33–4, 138
Zhuhai International circuit 31
Zolder circuit, Belgium 116, 145